What people are saying about

The Scars of Eden

There is absolutely no doubt. Our planet was visited by extraterrestrials in the deep past. Paul Wallis discusses the variety of views on the [question] of extraterrestrials. Exciting. I absolutely recommend *The Scars of Eden*.
Erich von Daniken

Paul Wallis' extraordinary works of investigation bring much humility and insight to one of the most contaminated fields of enquiry, 21st Century Ufology. History will recognise Paul's important works as a solid foundation in understanding humanity's origins.
Jaimie and **Aspasia Leonarder**

Fantastic book! A must read for all of those searching for answers! Paul proves yet again that human civilizations have been visited and influenced since the very beginning by visitors from the stars.
Matthew LaCroix

Also by this Author

Escaping from Eden
ISBN: 978 1 78904 387 7

The 10 Steps to Finding Your Soul-Mate – Guaranteed (results may vary!)
ISBN: 978 0 9873904 0 0

The Scars of Eden

Has humanity confused the idea of God
with memories of ET contact?

The Scars of Eden

Has humanity confused the idea of God
with memories of ET contact?

Paul Wallis

6TH
BOOKS

Winchester, UK
Washington, USA

JOHN HUNT PUBLISHING

First published by Sixth Books, 2021
Sixth Books is an imprint of John Hunt Publishing Ltd., No. 3 East St., Alresford,
Hampshire SO24 9EE, UK
office@jhpbooks.com
www.johnhuntpublishing.com
www.6th-books.com

For distributor details and how to order please visit the 'Ordering' section on our website.

Text copyright: Paul Wallis 2020

ISBN: 978 1 78904 852 0
978 1 78904 853 7 (ebook)
Library of Congress Control Number: 2021930336

A CIP catalogue record for this book is available from the British Library.

Design: Stuart Davies

UK: Printed and bound by CPI Group (UK) Ltd, Croydon, CR0 4YY
Printed in North America by CPI GPS partners

We operate a distinctive and ethical publishing philosophy in
all areas of our business, from our global network of authors to
production and worldwide distribution.

Contents

Acknowledgements

Thank You to the whole team at *John Hunt Publishing* for your amazing work and your bold investment in *The Scars of Eden* and its prequel *Escaping from Eden*. To Erich von Daniken, Ramon Zurcher, George Noory, Sean Stone, Maxim Makukov, Richard Dolan, Matthew LaCroix, Alan Stivelman, Juan Perez, Jaimie and Aspasia Leonarder, Kevin Shepherd, Etinosa Ewemade, Jeph Oro, Candid Rose, Steven and Evan Strong, Joshua, Audrey, and Barbara Lamb; thank you so much for our many conversations, on and off the air. I am indebted to you for all your encouragement and collaboration. I am also deeply appreciative of the courageous men and women who contact me every week with their personal experiences and encounters. You give me courage for the journey.

The encouragement of my family, Ruth, Evie, Ben and Caleb, whom I love more than anything, has been absolutely outstanding. This book has in many ways been a family effort. My love and thankfulness! Thanks to my parents Rodney and Brenda who first introduced me to this topic and to my brother Mark who enthused me along the way. My gratitude to Kofi and Patience, my parents-in-law who showed me the intersection of ancestral narrative and our family history. A special thank you to Anthony Barrett, paranormal researcher and founder of *The 5th Kind TV* for his friendship and collaboration. My profound appreciation and thanks to my friend, and publicist, Gavin (G.L.) Davies. His compassion and courage in this same field of study are second to none.

For the sake of people's privacy, I have altered some details to make anonymous any references to personal experiences not already in the public domain. Thank you for allowing me to do that. I dedicate this book to all those who have carried stories of contact in courageous silence, because of the taboo which surrounds this subject. I applaud you and, with you, I look forward to the day when our world is more ready to honour those whose experiences test and transform our understanding.

Contact me at www.paulanthonywallis.com

Introduction & Chapter One

What Happened to Paul?

Bath, England – 1985

I know what it is to be asleep and I know what it is to be awake. Right now, I am wide awake. I know exactly where I am and can see clearly what time it is. It is 2am. The glow from the streetlight outside, and an open gable window with the curtains drawn back mean that everything in my room is clearly lit. I just can't explain what I'm seeing.

I live alone in a top floor apartment in a chocolate box village on the outskirts of Bath. Being on my own in my apartment has never worried or unnerved me. I am not a fearful person. I love my independence. In fact, at this time of the night I am usually out on my own, exploring the dark hillsides, and the beautiful open spaces of Bath Golf Club, deep breathing in the moist midnight air. Usually I return home around 1:30am for a warming nightcap, and finally climb into bed around 2am. That's my routine. Being out and on my own in the middle of the night doesn't worry me in the least. Why would it? I am twenty years old. I'm indestructible.

For a good night's sleep, I leave my curtains open and the window ajar to permit a good draught of fresh air. Thanks to the well-positioned streetlight, the open window always bathes my bedroom in a soft light. But tonight, in the warm orange glow something is moving, and I don't understand what I am looking at.

I can see five things just beyond the end of my double bed. They are standing side by side between the foot of my bed and the gable window. The figures are short, grey, almost translucent, and moving just enough to show me that they are alive. But they are not human.

Knowing my voice will fail me, I hiss at them in a whisper. *"In the name of Jesus, get out!!"*

I am twenty years old, but I have pulled the bed covers over me like a two-year-old and I am shaking like a leaf. Suddenly, I am not indestructible. I am terrified.

Canberra, Australia – 2020

I have just stripped off for a shower when the bouncing tones of Skype panic me up the stairs in my bath towel. I really don't have time for an interruption right now. I'm going to be on the air in a matter of minutes and I need to be ready. Why is the call coming through now, when the interview isn't due for another half hour? Dripping and out of breath, I pick up.

"Joining us live all the way from Canberra, Australia is Paul Anthony Wallis, researcher of world mythologies and author of the controversial new book Escaping from Eden. *Paul, welcome to* Zone 51. *How are you today?"*

I do my best to slow my breathing and try and sound smooth and collected, *"Thanks, Tim and Jay. It's great to be with you today. I... I had in mind we were going to be talking in about half an hour!"*

"Yes," they laugh. *"The clocks' changing has thrown everybody out! So, Paul, your book is stirring up quite a hornets' nest of debate. Yet a couple of years back you were a regular pastor going about your work in the church. In fact, you were a senior churchman – an Archdeacon! What got you caught up in world mythologies and ancient aliens?"*

I sit at the end of my bed, wrapped in my towel, and thank my lucky stars this interview is audio only. We're quickly going to get into some deep territory and all my notes are waiting for me in the other room. Can I get through a two-hour interview without my notes? Is my middle-aged mind up to the challenge? I feel I've been fumbling more for my words lately, starting a sentence and then having to work hard to make it coherently to the end. (I'm guessing three kids and not enough sleep can do that to you.) I really don't want foggy memory to let my subject matter down and I'm feeling vulnerable sitting in my towel without my notes at hand. But it's not the first time I've been asked, and it is all about my own journey. It's about a sequence of discoveries that have upended my career, put my reputation on the line and changed the whole direction of my life. It is my own story. I can simply tell that. And so we begin.

We laugh about how an ultimate frisbee injury laid me up for weeks on end and gave me the time to get into some study while locked down in the shipping crate cabin at the end of my driveway. We talk about how I had been waiting for years to drill down into some of the logical and moral problems within the familiar stories of beginnings to be found in the book of Genesis. Anyone who has read those stories would know the kind of thing I mean. So, I run through the clues in the Genesis story which point to an even older narrative hidden in plain sight within the text.

Next, I walk Tim and Jay through the translation issues around various key words and show that the word the Bible often translates as "God" is really much better translated as "Powerful Ones". I explain,

"As soon as you make that translation switch the familiar stories of Genesis quickly reveal themselves for what they really are. They are the summary form of an even older body of stories – the Mesopotamian narratives of the ancient Sumerians, Babylonians, Akkadians and Assyrians. And those stories are not about God. They're about our distant ancestors bumping up against visitors from another planet – Sky People – extraterrestrial visitors who came and colonised our planet and interfered in our evolution to make us a more useful workforce."

I am now shivering under the cold of the aircon, having sat myself right under the vent. I try to steady my voice by talking louder. I don't want to come across all anxious and nervous when in reality I'm simply going into thermal shock.

"*Paul, after two thousand years of Christianity and more than three thousand years of Judaism, how likely is it that out of the blue, some random Australian researcher, yourself of course, could come along and say, 'Hey, everyone! You've got this all wrong! I've got it worked out! The Bible isn't really about God at all. It's all about aliens.' How credible is that?*"

This is a question I am ready for and as I jump into it I can feel my energy rising.

"*Well, Tim, if I really were the first person to make these claims then, yes, you would be quite right to raise an eyebrow or two. In reality, I'm not the first. If you go back to the very early days of Christianity you will find some very significant Church Fathers who were totally comfortable with the explanation of human origins that I am putting forward in* Escaping from Eden.*"*

"*I am talking about people like Justin Martyr, Clement of Alexandria, Origen and Marcion. And in their case they didn't get these ideas from the Mesopotamian stories. They got it from Plato.*"

"*You see, four hundred years before Jesus, Plato had already told the ancient world about the extraterrestrial beings who came and modified our ancestors. He called them 'children of God'. He didn't say what they were or where they were from, only that they came and adapted our ancestors to give us a higher capacity for consciousness and intelligence. Plato also wrote about some other company in the universe. He talked about others who are more advanced than we are. He said they live longer, they are more intelligent than us and they have an advanced knowledge of outer space. And, he said, they live on islands in the sky.*"

"*On top of all that Plato revealed that the Earth is a globe, floating in space, which every so often gets blasted by the movement of objects in space, triggering extinction level events. That's why, he said, we are not the first civilization on this planet. Every few thousand years or so, something happens to take us down to a virtual zero to start again.*"

"*Now those Church Fathers I mentioned, who were strongly into*

Plato, knew everything he taught about ET interventions and past civilizations and they had no problem endorsing Plato in the strongest terms. So that put all those topics on the table and made them part of the mainstream conversation in early Christianity."

"So, I am not the first on this territory. I am bringing back to the table things that in the beginning were part of mainstream conversation – and should be today."

I can feel my heart rate returning to normal and, although feeling a bit colder than I would prefer, I am beginning to get my groove on. I hope I'm not sounding too bullish. Talking about extraterrestrials is quite enough on its own to make people take a step back or get defensive. I am well aware that a hearer will automatically question the sanity of any speaker who trespasses into ET territory. I can't take offence at that because it's been a challenging change of mind for me in the last few years. It's a big ask to expect people to take that turn with me in the space of a two-hour interview.

Before I made that turn, I lived and worked for thirty-three years in the world of Christian ministry. For much of that time I worked as a kind of *"Church Doctor"*. Like many in that field, I had my fair share of paranormal experiences. Somehow, though, I was always able to interpret any anomalous encounters as Divine, human, demonic or psychiatric. Those were my boxes and everything I ever encountered, one way or another, could fit into one of those boxes. Somehow, my theology had never really found a place for other kinds of entity or other intelligent life in the universe. So it was a revelation for me to realise that our ancestors embraced far more subtle thinking on these questions than I had ever imagined.

"But, Paul, if in the distant past our ancestors knew about ET entities, how did all that knowledge get forgotten?"

It's a good question. It would seem a big thing to forget. So as swiftly as I can I lay out the dot point version of the answer:

- In the Ten Commandments it was commanded out of Judaism.
- In the C6th BCE it was edited out of the Hebrew Scriptures.
- In the C1st and C2nd CE it was excommunicated out of the early Church.
- It was illegalized from the Roman Empire by Emperor Theodosius in 381 CE and buried in the caves of the Nag Hammadi desert to protect it from obliteration.
- In the C15th it was confiscated and burned out of the libraries of Central and South America by the Spanish and Portuguese conquistadors.
- In the C16th its proponents were burned at the stake by the Roman Catholic Church.
- In more recent times President Truman suppressed it when he signed the National Security Act of 1947 and that policy of suppression has continued unabated right up until last year when a number of cats got let out of the bag by US Department of Defense.

"... So, Jay, to answer your question, you're right there has been a lot of forgetting and none of it by accident!"

I realise I have covered that ground rather quickly if you are not familiar with this history of official *"forgetting"*. But don't worry, in the next few chapters I will show you exactly how all that played out.

I will also show you how a living memory of ET interventions has survived this history suppression in the form of indigenous memory – what we call folklore or mythology. When digging for what our ancestors knew that didn't make it on to the news, our world mythologies and ancestral narratives are the most vital source. It is in these grassroots stories, passed from one generation to the next, that our ancient memories survive. In the pages that follow I will share with you my various notes on the subject and you will hear some stunning examples of *"prehistoric memory"* surfacing and resurfacing in cultures all around the world.

"*So, Paul, after thirty-three years in ministry you have boldly gone where few pastors have gone before! You must have ruffled a few feathers along the way. How many colleagues and friends have you lost through putting this material out there?*"

This is another good question! The fact is I have lost a small number. Honestly, though, as I have begun getting out there on George Noory's *Coast to Coast*, Sean Stone's *Buzzsaw 2020*, and countless other interview shows and podcasts, what has amazed me more than anything is the constant stream of letters, emails and messages that come to me from all kinds of people, all desperately relieved to hear a calm voice speaking into the airwaves on the topic of ET contact, ancient and modern. From week to week I hear from people of all ages and from every walk of life – scientists, engineers, nurses, teachers, police, defence personnel, researchers, therapists and pastors. Often those who reach out to me have experienced phenomena they don't understand and can't fit into conventional categories. I hear stories of close encounters, sightings of craft, interactions with non-human entities, even abductions. What humbles me the most is when from time to time I hear mature men in their sixties saying something like this:

"*I have told my wife and I've spoken to the person who was with me when the encounter happened. And I haven't told another living, breathing soul in the fifty years since.*"

That's how powerful the fear of shaming and ridicule can be. Yet all these decades after their experiences, my correspondents still need to find someone to talk to about it. They still need to process what it was that happened to them all those years ago.

As I have listened, I have come to feel even more passionately about the importance of breaking the taboo surrounding these subjects. Surely, we only impoverish our understanding of the universe if we embarrass into silence anybody with an anomalous experience to share. After all, the history of scientific discovery shows us that anomalies are our friends – if we want to understand reality more accurately. We just need to be willing to put our pre-packaged assumptions and beliefs back on the table for a moment and listen with an open ear to some accounts and experiences which may be puzzling, confronting and hard to understand.

"So, Paul, why does this matter to you so much? Wouldn't you be better off to keep your head down, and keep your speculations to yourself? What makes it important to you? Have you had some kind of a close encounter yourself?"

If you had asked me this question only a few months ago, I would have simply said, *"No."* Now, having compared notes with so many experiencers, I can feel something rumbling away in the back of my mind. There's a restless memory, itching to resurface. I just don't have a handle on it and find myself caught off guard by Tim's question. Doing my best to think on the spot, I stutter my way through an answer. Whatever my own experience, I can say that, for whatever reason, I share a deep fellow-feeling with all those who share their stories with me.

As Tim has correctly surmised, not all have welcomed my trespassing into this controversial topic. And yes, I may have lost a few friends. Commenting on the publication of *Escaping from Eden*, two of my comrades in ministry told me just recently, *"Paul, let's stay friends. But I won't be reading your book."*

Other friendships which were previously free and easy are perhaps not quite as free and easy as they were before. The fear of *"What has happened to Paul?"* or *"What if Paul undermines my faith?"* has kept some friends a degree more distant than in times past. Without a doubt, exploring ET territory can be somewhat isolating. With some at the fundamentalist end of the spectrum there is absolutely no possibility of a calm conversation. For those in that milieu I am a wolf in sheep's clothing. I am, *"Full of the pride of Lucifer,"* says one. *"Luring people to the pit of hell,"* says another. *"You need Jesus, man!"* comes one reply. *"Open your Bible and read it, idiot!"* comes another.

Naturally, I want to defend myself. I want to tell my detractors, *"No, you're wrong about me! I have been a believer since I was 17. I have not only read the Bible through countless times, but I served for fifteen years as a theological educator. I have trained pastors in the principles of hermeneutics [the principles of interpreting ancient texts – the Bible in particular]."*

I want to reply that I have designed training programs for pastors, served as a troubleshooter for churches, and as an Archdeacon in the Anglican Church in Australia. I really am a person of faith! For some, though, it is automatically impossible to acknowledge a person who accepts the possibility of ETs. Tony, my collaborator on *The 5th Kind TV,* jokes that I should merchandise a T-shirt that says, *"I lost them at 'Sky People!'"*

"So yes, Tim, I have to admit there is a cost to exploring this kind of territory. But I do it because it's my journey. I want to know the truth. And as an addictive writer I just have to share that journey with others."

I sign off from Tim and Jay's show, still sitting at the end of my bed in my bath towel. I think I did OK without my notes, which are still sitting undisturbed, patiently waiting for me in the other room. My hosts were right about how outlandish my claims sound on a first hearing and, generally, I am happy to take it nice and easy as I explain to anyone who's interested the steps that have taken me from the safe, predictable world of ministry and on to the world stage on such a controversial topic.

When I published a book with the confronting subtitle, *"Does Genesis teach that the human race was created by God or engineered by ETs"*, I knew I was firmly nailing my colours to the mast. That ship has taken me a long way. In these pages I will share with you something of the amazing voyage that has followed.

Together we will travel around the world – to Greece, Argentina, England, Wales, Scotland and Ireland; to Italy,

America, Peru, Kenya, Ghana, South America, the Philippines, India and Australia. We will sit at the feet of scientists and sages, scholars and saints to spur us on in our exploration. We will go to the Fesenkov Astrophysical Institute and the al-Farabi Kazakh National University in Kazakhstan and learn of incredible findings in the world of DNA from world-class researcher Maxim Makukov. We will take some time on America's East and West coasts to hear from Psychologists John Mack and Barbara Lamb. We will delve more deeply into the abduction narratives of ancient indigenous narratives in the company of Aboriginal Elder Shane Mortimer. Hebrew scholar and translator for the Vatican, Mauro Biglino will point us to key historic words which have been hidden and suppressed through centuries of questionable translation.

In Argentina we will meet filmmaker Alan Stivelman, and in America the eminent UFOlogist Richard Dolan. In the company of Anthony Barrett on *The 5th Kind TV* we will introduce ourselves to a body of deeply challenging contemporary research. With anthropologist Maria Scholten we will see the evidence of a longer and more interesting story of civilization than the one you and I learned in school. And along the way experiencers in Argentina, Ghana, America and Australia will add their personal stories to encourage us on our journey together and spur us on to keep exploring.

By the time you and I reach the end of our journey together, I am also hoping that the writer who first challenged me as an eleven-year-old boy to ask bigger questions might make an appearance for a personal conversation. I wonder what more he might have learned in the decades since his story first intersected with mine. I am talking about the author of *Chariots of the Gods*, Erich von Daniken.

The white rabbits who will meet us in the next chapter and lead us into this incredible rabbit warren are members of my own family. I will introduce you in the next chapter where we will descend the first dip of this roller-coaster and, let me warn you, you will need to hold on to your hat.

Before we accelerate down that first dip, I want to take a moment to thank you personally for picking up this book and agreeing to share this journey with me. It is my pleasure to have your company. Because, for me, as an addictive writer, it's about sharing the experience, and putting into your hands the book I wish someone had given me many years ago! When I first set pen to paper to map out what I am now going to share with you, I actually had no idea what forgotten treasures would resurface in the process – least of all, buried in the depths of my own memory.

Chapter Two

The Secret Commonwealth

Canberra, Australia – 2020

What company must we be keeping on planet Earth to generate the incredible stories I have been uncovering? What traumas must lie in our ancestral past to have scarred our collective memory with such disturbing mythologies? And how much of what our ancestors have reported bears any relationship to the real world of the C21st?

My burning questions have led me to publishing two books which I fully realise put my reputation on the line and lay me open to ridicule. As I bring *Escaping from Eden* and now *The Scars of Eden* into the light of public scrutiny, I have to be willing for people to judge – both before and after reading. However, for me, coming from a background of thirty-three years in ministry there is another layer of challenge. A pastor should be a reassuring figure, and some consider even the mention of prehistoric ET contact to be an assault on their theology – let alone ET presence today, let alone contact, let alone abductions. Some religious believers respond as if I must be an agent of the devil to raise questions which, on first glance, appear to be outside the ambit of mainstream religion. So, you can probably imagine that it was with some trepidation that I prepared to sit down with my parents-in-law to let them know the topic of my forthcoming book.

My wife Ruth's mum and dad are devout Christian believers and their spiritual tradition is Baptist and Pentecostal. I had no idea how they would react to what some would regard as an attack on orthodox Christianity. On a visit to our place in Canberra one weekend, we sat down together after a Sunday feast of Ghanaian cuisine and, as disarmingly as I could, I laid out my thesis concerning the Bible's narrative of extraterrestrial interventions in our evolution, as well as the stories the Bible shares with other world mythologies concerning a pattern of abduction and hybridization through the ages.

My parents-in-law listened quietly and politely. Then with an animation that completely took me by surprise, they leant forward and said, *"Paul, we already know this story. People in Ghana know about this. Our story goes back for generations. And we know people who had this thing that you are talking about happen in their own family…"*

So it was that I came to hear a personal story from my own family that I previously had no idea about. The events it relates are nearly forty years old. I will now pass the story on to you, but please permit me to leave some details blank for the sake of protecting our friends' privacy.

Anloga, Keta District, Volta Region, Ghana, West Africa
October 1984

Dazed and anxious, walking on the beach of Anloga is a young woman. Akua is twenty-six. As she gets her bearings, she begins to make her way back to the family house a few blocks away. When Akua walks through the door her family is stunned. They gather around her with tears of joy and confusion. She has been missing for three years.

As they scurry around to make her comfortable and help her to settle, Akua's family cannot help themselves plying her with questions. Where did she go? What happened to her? Why has she not contacted them in all this time?

Gradually, sadly, Akua reveals her story. She was taken from the beach in 1981. The place where she was taken was somewhere she didn't know; somewhere hidden and secure, and free of any devices by which she could get messages to the outside world. Her captors held her against her will and forced her to bear children. Exactly how they have returned her to

Anloga beach is something of a mystery to her. Concerning that journey, her mind is a fog.

Akua's account is deeply disturbing for her loving mother and father to hear, and their immediate concern is to help her begin to feel safe and secure again in her family home. They do their best to reassure Akua that all is now well, and that they will do everything in their power to help her forget and move on with her life.

But as the weeks go by it becomes clearer that Akua has kept something secret about her abduction. It is a secret that has left a cloud of anxiety perpetually darkening her young face. After long pleading from her mother, Akua finally agrees to tell them what it is that still terrifies her concerning her ordeal. What she has to say is like nothing Akua's mother has ever expected to hear. Between tears, Akua whispers her secret.

"My captors," she hesitates, "were Mami Wata. They were not human."

Canberra, Australia – 2019

When Akua returned to her home in Ghana three years after her vanishing, her family prepared for the unfolding of a horrible story – a forced marriage, a secret elopement, kidnapping or slavery. Such things are hardly unknown. However, the account their daughter brought them was far and away from anything they had expected. Yet when Akua spoke the words, "Mami Wata," they knew what she was talking about.

For centuries Ghanaians have told stories of Mami Wata abductions. The history of this narrative is so ancient that its roots have become lost in the mists of time. On my own account, before that Sunday lunch with my parents-in-law, I had known nothing of the Mami Wata tradition. As I sat and listened to the ancestral wisdom of my own family, I learned that accounts of Mami Wata are not necessarily stories of ET contact. The Mami Wata people are generally identified as highly intelligent, extraordinarily beautiful, and humanoid in form. They are often very compelling, with promises of advancing the intelligence, health, and prosperity of those they take. And they operate, so the tradition goes, out of underwater bases.

This was a story like nothing I had heard before. My Ghanaian family are Pentecostals, Catholics and Methodists. I had not realised how close I was to another, more indigenous tradition.

In some places Mami Wata has morphed into a kind of religious or spiritual practice, not unlike the local rituals of appeasing territorial spirits which can be found in other places. The religious version of the story sees Mami Wata as a female entity with power over the seas. Not necessarily malevolent, but so superior in power that you would never want to get on the wrong side of her.

Mami Wata is a tradition of abduction narratives which stretches back in history for hundreds, maybe thousands of years. The Mami Wata reports relate more often to males being taken from near the beach, though abductions of females such as Akua are part of the narrative too. The purpose of these abductions, so the tradition tells us, is a program of hybridization. For some reason, so the mythology goes, the Mami Wata people wanted to create a new lineage of Mami Wata people, modified with human DNA.

I realise I have taken us quickly into the deep end. Without a doubt, abduction and hybridization probably constitute the hardest to swallow of all aspects of ancient ET claims. Yet in the panoply of world mythology and ancestral memory they are possibly the most widely recurring themes.

My earliest awareness of ancient hybridization narratives came from the Bible. It appears in the pages of Genesis 6 when the *"benei elohim"* (*Ones like the Powerful Ones*) came down to planet Earth to take human females. Their interbreeding produced a lineage called the Nephilim, who were the giants of legend. Both the author of Genesis 6 in the Hebrew Canon and the author of the letter of Jude in the New Testament assume that their readers are familiar with this piece of history as told in another famous text – the Book of Enoch – a text to be found today in the Ethiopian Orthodox canon of the Bible.

The Book of Enoch describes the abductions referenced in Genesis 6 in even greater detail. It names those doing the abducting as *"Watchers"* and describes how they had transgressed fundamental ethics governing relationships with the human race, arriving from somewhere among the stars to begin their program of producing hybrid offspring.

Josephus, the Jewish historian of the C1st CE appeals to this scriptural episode, to explain the presence of giant human beings through the annals of history and at the time of his writing. Josephus understood the Jewish tradition to be only one iteration of this legend of human hybridization with advanced beings. Furthermore, he identifies the Ancient Greek legends as vehicles of the same memory. The Greek name for these hybrids was Titans.

The Greek legends and the Hebrew mythologies are not the same, but both report external interventions in our evolution as a species. Both outline a program of hybridization which was not only traumatic for our ancestors, but which deeply divided the community of superior beings associated with *"project Earth"* at that time. These strange notes sound again in ancient Indian, Norse and Celtic narratives. Ancient Egyptian and Native American traditions also hint at something similar.

The name *"Mami Wati"* for the ancient Ghanaian version of the story is a new name. It dates from the British occupation of the Gold Coast, which shortly after its liberation in 1950 took the African name of Ghana. However, by other names, the Mami Wata tradition is far older and wider. Reports of Mami Wata abductions are told from Kenya on Africa's east coast, down to the southernmost tip of South Africa, all along Africa's Western seaboard and into the Caribbean, Haiti, Brazil and

Cuba. Curious parallels can be found in stories from Alaska and the Philippines.

For instance, the Luo people of Kenya have curated the story of Nyamgodho Wuod Ombare. It tells of a man, Nyamgodhu, and it tells us quite a lot about him. Nyamgodho was a Bantu man of the Waturi tribe, who lived in the C14th and C15th in the village of Nyandiwa, in the Suba district of Kenya, on the shores of Nam Lolwe (Lake Victoria). One morning, so the story goes, he found a strange-looking woman caught up in one of his fish nets. The fisherman agrees to marry the woman and live with her because she persuades him that she can bring him unimagined wisdom and prosperity. Which she does. But she gives Nyamgodhu a condition, that he must never speak of her true place of origin in the water. Some familiar notes are sounding.

In Kenya Mami Wata are identified with the *Jini* – the Swahili equivalent of the *Jinn* of the Arabic peoples. Moslem and Swahili people on the Kenyan coast also associate these abduction stories with entities called the Mahurani. Like the Mami Wata tradition, the Mahurani tradition is for some a ghost-like, spiritual phenomenon, while for others a more concrete story is told – a story in which a Kenyan man or woman encounters a strange person who is tall, white and unbelievably beautiful.

These tall white visitors – the Mahurani – appear to individuals by teleportation. Their visits begin in the contactee's childhood and continue into adulthood. On later visits, the Mahurani take their contactees to bases underneath the waters of the Indian Ocean. Here the contactees are used to produce hybrid children, who from time to time, they are allowed to see. However, the children are kept by the Mahurani who, after each abduction episode, return their contactees to their Earthly homes, unharmed and often healthier than when they left.

The families and religious communities of Kenyans who report these experiences are often confused and upset when their loved ones return, speaking affectionately of their beautiful, tall, white captors. Indeed, in Kenya such testimonies are often regarded as evidence of mental illness or demonization. This kind of dissonance reminds us that these kinds of abduction narratives are part of a tradition that stands separate from and previous to the official mainstream religions. The abduction traditions of Kenya find their source neither in Christianity nor in Islam. They exist within the unofficial, indigenous knowledge of the people, carried orally, and puzzled over, from one generation to the next.

Elements of Mami Wata or Mahurani folklore repeat in the indigenous Filipino story of the *Engkantos*. The *Engkantos* are shape-shifting entities, associated with the water. Their humanoid appearance is almost identical to our own. They could almost pass for regular Filipinos, apparently, save for their unusual skin-colour and the lack of a philtrum on the upper lip. The *Engkantos* are known for abducting human beings who get too close.

The Filipino word *diwatas* is another word that invokes ancestral knowledge of non-human entities, who abduct human beings. The *diwatas* – so the tradition goes – like to maintain a covert presence and don't take too kindly to their cover being blown. For that reason, people often refer to these entities obliquely in whispered conversations as *"dili ingon nato"*, which means *"those unlike us"* or *"lamangdagat"* which means *"those who dwell under the sea"*. These strange *diwatas* are physical beings, similar enough to humans to be able to procreate with those they take in order to create hybrid people. Does any of this sound familiar?

The very fact that such specialised language has been developed to carry this story reveals how important it has been to memorialize this information both in Filipino culture and beyond. The word *diwatas* suggests that the Filipino version of the story may have originated in India. This is because *diwatas* is derived from an ancient Sanskrit word *devata* which means a *"god"* or *"superior being"*. In a similar vein the vodou tradition of Haiti speaks of *Simbi Nan Dlo*, the spirit of the water. African slaves brought the tradition to the Caribbean with the name *Yemoja*.

Some of the words that belong to these traditions are incredibly intriguing. They have associations with spirits and serpents and, just like the word *elohim* in the Hebrew stories, they can often be translated to mean either a god or a demon. This 180 degree ambiguity is another reminder that we are dealing with ancient stories, stories that long predate the familiar binaries of good and bad, light and dark, that came with later importations of organized religion. With these ancient abduction stories, we are really in at the deep end.

Now some might want to suggest that I am taking these stories too literally. Might these ancestral stories be carrying the memory not of alien species but of foreigners? Could these mythologies of abductions be metaphors for the experience of the slave trade. The coastal homes of the mythologies in question would all have experienced a measure of slave trafficking.

For me this explanation holds little water. The patterns that repeat in the stories like those of the Mami Wata, Mahurani, Yemoja, Jini, Jinn, Engkantos, Diwatas, Dili Ingon Nato, and Lamangdagat are of young people being taken from near the water, held for only a short time in an underwater base, and then returned unharmed, but having been exploited for hybridization. Suffice it to say those are not the patterns of human slavery. It is a different pattern.

As I researched these indigenous mythologies, and began joining the dots from continent to continent, I found more and more cultures offering similar narratives. To my surprise I soon learned that the same phenomena of abduction and hybridization, along with specialised language for it, exist within European narrative traditions. In the nineteenth century folklorist Edwin Hartland argued that behind the *"fay"* and *"faerie"* traditions of Europe's Celtic peoples lay a real world phenomenon. With great courage he put forward the view that there is in Celtic folklore the memorialisation of a flesh and blood program of hybridization between human beings and another human-like presence occupying planet Earth, going back hundreds and maybe even thousands of years.

Two centuries before, Reverend Robert Kirk published his own research into the indigenous narratives of the Scottish Highlands. His book is still in print. The picture he pieced together described a covert layer of non-human interference or even governance over human affairs, which he referred to as *"The Secret Commonwealth"*. It was a shocking conclusion for people to hear from a Scottish Presbyterian Minister.

The Irish iteration reinforces the notion that this other presence wishes to keep itself secret. It echoes the curious detail from the Filipino tradition of not naming these other entities directly, referring to them instead with epithets such as *"those who dwell under the water"* and *"those unlike us"*. The whispered Irish epithets are, *"The Good People"* or *"The Noble People"* or *"The Other Crowd"*.

A Welsh telling can be found within the accounts of the *"Tylwith Teg"*. Though often taken as fairy tale of the fictional kind, some familiar notes resound. The eminent American anthropologist Walter Yeeling Evans-Wentz noted an unusual story within the *Tylwith Teg* canon, which described the existence of human-like beings who had the appearance of beautiful young women. They would meet young men near the water's edge and then entice them to follow them into their underwater communities to live with them as their *"husbands"*. Should any of the husbands choose to leave, they would have to return alone.

Within the Irish canon we can find stories of abduction for hybridization which would be indistinguishable from the testimony of twenty-first century claimants of abduction or close encounters with extraterrestrial beings – right down to the shape of the craft, the shape of the alien beings' faces and their speechless mode of communication. All that differs is the vocabulary used. Absent of terms such as extraterrestrial, alien, small grey or interdimensional, Irish stories tell the tale of *"faeries"* and *"little people"*. So perhaps reporter Richard Carleton's reference to being *"off with the pixies"* was not so wide of the mark after all – even if it was not meant kindly.

As I reflect on the geographical spread of these narratives, my mind goes back to the first time I heard a story of abduction and hybridization. It was far away from Scotland, Ireland and Wales, far away from Africa and the Caribbean, in a place whose legends have inspired and fascinated people for thousands of years.

Knossos on the island of Crete
July 1985

It is the era of terrorism. Around the world bombings and hijackings are regularly punctuating the international conflicts of the day. My dad's work is in combatting the impact of these troubles on our ability to travel the world. It is as a thank you for his work that our family is in Greece this summer, exploring the unique treasures that this beautiful country has to offer. We are touring the Greek islands as the guest of the Greek shipping magnate Andreas Potomianos, and the amazing world of antiquity is a revelation for me.

Crete is a beautiful island, and Knossos is its most significant megalithic site, dating from the Bronze Age. It is an inspiring feeling to stand in the place of an ancient civilization. In its heyday, Knossos was home to the Minoan people. Their culture thrived here between 3000 and 1000 BCE. The Minoans were an advanced, maritime culture, whose cities were impressive multistorey environments, with exquisite architecture, plumbing and airflow management. Their advancement in technology catches my attention. So, I put a question to our guide.

"Who exactly were the Minoans that they were so advanced? Where exactly did they come from?"

The answer comes in the form of a story I have never heard before.

"The Minoan culture appeared five thousand years ago," the guide begins. *"They were the people of the great ruler, Minos. Minos was a powerful man – although not really a man. He was a hybrid. The mother of Minos was a very beautiful woman. She was a daughter of Agenor, a king of the Phoenicians. One day she was walking on the beach with her young friends. Her beauty drew the attention of the ruler of the gods, whose name was Zeus. He instantly decided that he must have her."*

"Then, as if from nowhere, a beautiful and gentle bull appears on the beach beside her. The bull captivates the young woman with his animal beauty and gentle spirit. She strokes the animal and pets it. Something compels her to climb on to its back whereupon the bull gallops into the waves, taking her far from her home and family in Phoenicia, all the way here to the island of Crete. It is here, on this very island, that the young girl discovers that her captor is not a bull but a handsome god. You know what they say about Greek gods? They are very, very good looking! Zeus seduces the young woman and when the time comes she produces three sons – Rhadamanthys, Sarpedon and Minos. Minos became the ruler of the Minoan culture. His mother's name was Europa. The entire continent of Europe is named after her."

I had never heard this story before.

"So she was abducted?"

"Yes, exactly! Europa was abducted by a handsome Greek god!"

"And what happened next?"

"Europa was an unmarried mother with three children – which could have been difficult. But fortunately for her she was a very beautiful woman. She knew the King of Crete, and he loved her so

much that he married her and was willing to adopt her three sons as his own. And so they all lived a happy life. All except for Europa's first lover, Zeus, who was killed in a battle."

I was open-mouthed. This was something I had never learned in school. The whole of Europe is named after an abductee – a young lady who was taken from the beach. I had always thought that Zeus was an ancient Greek almighty God. This legend paints a different picture. Zeus is flesh and blood; something similar enough to a human being to hybridize with a human. Yes, he is powerful. He is able to manipulate the consciousness of Europa so as to cloak his true appearance. But he is not all-powerful. He is not even immortal. In the end someone kills him in a battle. This most famous of the *"gods"*, according to this version of events, was an advanced being who used his superior power to abduct and hybridize. The very name of Europe embodies a story of prehistoric hybridization. My time in Greece has sown a seed of curiosity that will grow for many years to come.

On our final evening in Athens we settle into our cabins on our ship, the *World Renaissance*, then the flagship of the Epirotiki cruise line. It is as we change into more formal attire for dinner that I notice something odd. Standing bare-chested in front of the bathroom mirror I see something that does not belong. Three raised lines on my belly, just below and to one side of the navel. The lines are about eight centimetres long with a pinprick effect that looks uncannily like the scar of a multiple vaccination. What is it? I haven't had an inoculation in my belly. It's hardly something I would forget. Being a mere twenty years old, I take myself into the next cabin for my mum to provide a more expert diagnosis.

"It's probably just some kind of rash," she says offhandedly.

But it doesn't look like a rash to me. I don't know what it is. And how long has it been there? Whatever it is, I put my mind on other things and forget about it. Indeed, it will be many years before I even remember the thing that happened in 1985.

Today I wonder more than ever how a testimony so widespread, carried by so many diverse cultures and traditions could remain so unknown? In the West of the twenty-first century when people like our friend Akua in Ghana claim to have experienced an abduction at the hands of something non-human, we startle as if we have never heard the like. As I journey through Greece, South Africa, Ghana to Kenya, Cuba, Brazil, the Caribbean, the Philippines, India, Scotland, Ireland and Wales, it is becoming clear to me that *"the like"* has in fact been told to us since time immemorial. We just haven't taken it seriously.

Western culture today appears unable to take these kinds of reports seriously. What we now call *"close encounters of the fourth kind"* must be categorised as fable or fiction, metaphor, or madness. The one thing we will not do with a person claiming a flesh and blood abduction experience is listen with respect. At least that is what I thought. Then one day in 2018, the revered journalist and national icon, Ita Buttrose, chair of Australia's ABC, spoke some words on national television that made my jaw drop.

Chapter Three

Look at the sky!

Canberra, Australia – 2018

"Jane's first experience of aliens – greys – happened when she was a toddler."

I have lived in Australia for more than twenty years and for all that time I have been aware of Ita Buttrose. She is a national icon, a celebrated journalist and editor. As a woman in national journalism, as the pioneering editor of *Cleo* magazine and as a popular figure in the Australian media Ita has become part of everyone's wider family. In 2018 she was appointed as the new chair of the Australian ABC at a time when the independence of Australian journalism has needed a hero and a champion. I say all that to let you know, if you're not familiar, that Ita Buttrose is an absolute pillar of Australian society. That's why my jaw drops as I hear her speak calmly, respectfully and without judgement to a lady by the name of Jane Pooley on a topic far from Australian mainstream culture – alien abduction.

Jane Pooley is a retired nurse and mother of three from central New South Wales. In the words of Ita Buttrose, she is a *"mild and assuming woman."* Her claim is that she is a lifelong experiencer of extraterrestrial visitations. She also claims to have been used by her visitors as part of a program of human-ET hybridization, and to have produced two hybrid children, who were taken from her mid-term.

Before the interview on national television on *Studio 10*, Jane presents the show's team with medical documentation evidencing her pregnancies, along with other objective indicators of her story. The team then invites her to take a polygraph test to be quizzed on the key elements of her story, to which Jane readily agrees.

"Did you meet an alien?"

"Do you have part-alien children?"

"Have you been part of an alien breeding program?"

Jane underwent the polygraph test three times and the result, says Ita, *"No deception was indicated."*

The piece on *Studio 10* presents no more than the bare bones of Jane Pooley's account. It makes no reference to the corroborating evidence that Jane has presented in terms of medical papers. Neither does it refer to the support of other eyewitnesses. No doubt it was easier to get the show's producers to agree to a *"You decide..."* type piece, than a piece that says, *"The evidence clearly points to an astonishing conclusion!"* I can understand that. But what strikes me is that this is the first time I have seen any reference in Australian mainstream media to the kind of things I have been learning from my Ghanaian family, and from mythologies around the world concerning a widespread and ancient report of forced human hybridization with an extraterrestrial presence on Earth. To see a figure of the standing and credibility of Ita Buttrose take the story on, albeit non-committedly, but calmly and with respect – is astonishing. Something in our culture is shifting. In fact, something has shifted a long way. Fifty years ago, it was a different story.

Westall, Clayton, Melbourne, Australia
11:00am – April 6th, 1966

A young boy bursts open the doors of the science room on to the school playground.

"Mr Greenwood! Mr Greenwood! There's things in the sky! There's flying saucers in the sky!"

Until now it has been an ordinary day for the children of Westall Primary and Western High school. Now the corridors are full of children and staff racing on to the school playground. As Mr Greenwood the science teacher runs out of the building another student shouts out.

"Look at the sky! Look at the sky!"

Children are running, screaming, shouting and crying. One girl even faints as three silver, saucer-shaped craft hover over adjacent the schools. The chemistry teacher, Mrs Sandleby, grabs her camera and starts taking photographs.

After ten minutes one of the craft appears to put down on the Grange – a lightly wooded area immediately adjacent to the school grounds. The boys and girls charge down to the fence to catch a closer look at the flying saucer. Two of the girls, Jacquie and Tanya, jump over the fence and Tanya runs ahead through the pine trees, hoping to touch the craft. Within seconds Tanya reappears, running back towards the fence, screaming and hysterical. Minutes later she has been taken to hospital.

In the skies above, five small aircraft now appear, and for about twenty minutes they manoeuvre around the other two flying saucers, trying to get a closer look at whatever these unusual craft might be. As more children approach the Grange, the landed craft rises to an altitude of about twelve feet, leaving marks and a circular impression in the ground below. It then rotates on to its side and retreats into the sky at phenomenal speed. As it does so, the five small planes do their best to track it but are simply left standing.

Within thirty minutes of the saucers' arrival military jeeps arrive at the school, carrying army personnel in fatigues. Cars and vans now converge on the schools, bringing police, a high-ranking government official, and press reporters. At the school fence a twelve-year-old student, Joy Clarke, is talking to an interviewer from *Channel 9*, explaining what everyone has just seen.

From out of nowhere a man in a dark blue suit steps up to Joy and orders her to stop talking and go inside. The man then commands the film crew to stop filming. Inside the school another man in a dark blue suit has accosted the chemistry teacher who has been taking photographs of the three craft.

After a heated exchange the man wrenches the camera from the teacher's hands and takes it away – never to be seen again.

The same day, that very afternoon, the schoolchildren find themselves herded into a special assembly. Their principal has an important announcement to make. He says,

"Nothing happened this morning. Any child who claims that something did happen will be punished."

The stunned children sit dutifully as the principal makes clear to staff and students that what they all witnessed, what had sent children into a panic, what within thirty minutes had brought the police, the air force and the army out to respond was in fact a weather balloon. Any student offering a different explanation will be put in detention.

The next day Jacquie – one of the two to jump the fence into the Grange the day before, ran round the house of her friend Tanya – the one who had been taken to hospital. She wanted to know if Tanya was feeling better. To her utter amazement the door of her friend's house was opened by a stern, English-speaking woman. Tanya's parents spoke Yugoslavian. The woman insisted, *"Your friend Tanya doesn't live here and has never lived here."*

Jacquie never saw her friend Tanya again.

Years later Jacquie learned that Tanya had been removed from the school overnight and put into a convent. More than a hundred students have been willing to speak publicly about what happened. But to this day Jacquie's friend will not speak of the incident nor the part of her life which followed.

This is the Westall Incident. It is the biggest mass sighting of UFOs in Australian history. More than two hundred students witnessed it, along with teaching staff, local workers and residents. Army, Air Force, Police and Press all responded to the incident. And yet to this day the official government report is that nothing happened. There was no incident.

Ten years ago, Australian researcher, Shane Ryan, conducted

an investigation into the case and published his findings. Not long after, Shane received a phone call from a brother and a sister, now in their fifties. They were the son and daughter of the government official who had been despatched to Westall on that fateful day. At that time, their father was a senior public servant in the Department of Supply.

When Shane sat down with the man's son and daughter in 2010, a sad story unfolded. They told him that their father had died, still a young man, only four years after the incident. After what he had witnessed at Westall, he had been determined to get to the truth of what had really happened on that April day in suburban Melbourne. But, evidently, their father's determination had clashed with strict instructions from on high to stop digging into a story that the government wished to keep under wraps. The pressures – and no doubt threats – which issued from that clash were, in the view of his son and daughter, the cause of his untimely death.

Shane Ryan then took a trip to the *Channel 9* archive to take a look at their film reportage of the incident, including the interview footage which had got two students into detention all those years ago. Scanning the archive of dusty tins, Shane was thrilled to locate the tin labelled April 6th, 1966. His excitement quickly turned to dismay as the container revealed that at some point in the thirty-four years since, someone has seen fit to remove the contents.

Today, as far as the official government narratives go, there was no Westall Incident on April 6th, 1966. Despite the military, air force, police and government presence involved, apparently no government agency has retained any record of an event ever having happened on that day. As yet, Australian government agencies have not even found a way to say, *"Yes, there was an incident. We just don't know what it was."* Just nothing. Silence.

What was it in 1966 that compelled the government to this policy of silence? And what makes the incident so sensitive that it still cannot be acknowledged, more than fifty years later? You have to wonder precisely what cat would be out of the bag if the Westall Incident were to be acknowledged at an official level?

But while our government policy has not shifted, it seems to me our culture slowly has. The children of Westall, silenced and threatened in 1966, are now allowed to speak openly – and even go on national television to testify to the biggest UFO mass-sighting in Australian history – and to do so without consequence.

The Westall Incident powerfully illustrates why the mythologies and indigenous narratives of the world are so valuable. It is in these unofficial stories, in our folk memory, that our cultures carry the memory of moments and events unacknowledged by our political authorities.

Seeing and hearing the witnesses of Westall speaking freely in the twenty-first century encourages me to believe that our culture may finally be a degree more ready to listen than it was in a previous generation. Yet, even with more people more willing to listen, the fear of ridicule remains a powerful lid on our sharing of memory to this day.

Some of those who have contacted me with their own stories of strange encounters have taken more than fifty years to reach out to anyone other than their spouse to tell of their experience. A great many have been kept silent by the fear of ridicule. Of all the experiencers I have spoken with, among those most powerfully affected was an Argentinian gaucho – a farmer whose life was changed at the age of twelve by a phenomenon he could not explain.

Acevedo, Pergamino, Argentina
1978

Juan Perez is out on his horse, checking on the herd. For Juan, this is a familiar part of his duties on the farm. But today he will encounter something unknown. The encounter will change his life. He is twelve years old.

The field stretches ahead of him towards the horizon which today, as never before, is hidden behind a dense cloud, which slowly advances towards him. As Juan enters the fog, the contours of something unfamiliar begin to take shape. What is it? Is it some kind of vehicle? Some kind of workers' hut? Unafraid, Juan examines the unfamiliar structure. He tethers his horse to it and quietly, inquisitively, he climbs up a kind of ladder to peer inside an open door. What he sees are two figures, working inside. One is taller, one shorter. There is something almost robotic about both figures. The taller figure then turns and approaches Juan. It is not human.

That day, Juan is one of three local witnesses who, in that small Argentinian district, report a close encounter with what appears to be an ET craft. The other two witnesses have merely seen a craft and managed to escape its attentions. Juan's encounter on the other hand will change his life forever.

My first image of Juan was in footage of him as a young man of eighteen. As the guest of a UFO congress in Argentina, he had been given the opportunity to share his story in front of a small, friendly audience. Sitting behind the microphone, he gave the appearance of an intelligent, confident young man. His first few words indicated a grounded and eloquent person. Yet only a few words into his presentation he became choked with anxiety and emotion and buried his head in his hands, mumbling, *"You won't believe me."*

Watching the same footage in Argentina was the film-maker Alan Stivelman. Alan's spontaneous response to this moment on film was to stand up and announce, *"I have to meet that man."* And meet they did.

On their first meeting Juan was still shut down, convinced that no one would believe what he recalled from 1978. In the years that followed the incident he has isolated himself, secreting himself away on his rural farm. Juan's self-imposed exile had continued for thirty-five years. Gradually, gradually Alan won Juan's confidence and slowly Juan began to open up about what it was that happened to him all those years ago. Alan re-introduced Juan to an eminent researcher in Physics and UFO phenomena, Jacques Vallee. He sat with Juan for sessions of hypnotherapy as psychiatrist Nestor Berlanda helped him to uncover some lost memories. He then journeyed with Juan to his ancestral home in Paraguay to learn the indigenous narratives of his ancestors. After four years of sharing the long journey of healing, Alan and Juan have made a beautiful film about the process. I hope I have told you just enough to whet your appetite to watch the movie – *Witness of Another World*.

Acevedo, Pergamino, Argentina
January 2020

The frightened, broken down boy of eighteen is now a man of fifty-three. Standing on his farm in Acevedo in rural Argentina, and fielding my questions, Juan appears healthy, looking relaxed and confident. When Alan began to film Juan for the movie, he was forty-nine. At that point, sadness and fear still filled the gaucho's face and he spoke haltingly of his close encounter and its aftermath. Today as we talk together, the change in Juan's demeanour is amazing, and a joy to see.

"The film changed my life," he begins. *"It healed my fear. Until that moment I had something that I could not tell anyone. I was ridiculed through my whole life. Until I turned 47 years old, I didn't know how to tell my story. Even remembering what happened to me made me afraid."*

"It is difficult to see something unknown – and even harder to explain it. People just don't believe you. Thanks to the success of the movie, people have begun listening to me. And that has helped me to open up to talk about these things."

As a teenager Juan tried to speak of his encounter and its aftermath but found he was made a figure of fun. That is why he chose to move further into the country to isolate himself on a farm for the next thirty-four years. And for all those years PTSD continued to afflict him. I ask Juan about the psychic phenomena which followed his close encounter.

"Between the ages of fourteen and fifteen," he says, *"I would have dreams and those dreams would come true. But when I talked about it, people refused to believe me. I have been spat on. People have refused to acknowledge me or shake my hand. I suffered a lot of discrimination. Jacques Vallee was one of the first people to listen to my story. The work of Dr Berlanda also helped me to understand what was happening to me."*

Dr Berlanda's regression work with Juan ultimately surfaced the buried memories of more than one encounter with the beings he first witnessed on that day in 1978. Even though Juan's memories did not include any kind of violence towards him, the very idea that ET beings had taken him to another place – whether physically or mentally – was a deeply disturbing realisation.

"How?" he asks. *"I don't understand that. Experiencing it is very traumatic. Only a person who has experienced such an abduction can really know what I am talking about. How could they have that kind of control over us? And why did this happen to me?"*

In addition to the memories, the strange phenomenon of precognition in dreams, and his symptoms of PTSD, Juan was left with another legacy from his encounter – a strange scar on his upper arm – three raised lines, each about 8cm in length, where one of the beings had touched him. These phenomena at least gave Juan some reassurance that something objective had indeed happened to him. Four decades after the incident, as he worked with Alan on the film, another layer of corroboration surfaced.

Ever so reluctantly Juan's mother confided with her son that she herself had experienced a terrifying close encounter in her youth. Juan's co-witness, his horse, had been startled by their bizarre encounter and died the following day. His mother's co-witness had been her dog. It too became agitated and anxious. Sadly, the dog was taken by the strange visitors and never returned. In the years that followed the encounter, like Juan, his mother too had experienced precognitive dreams.

Juan's mother believed that it was because of her contact with extraterrestrial beings that Juan had received the unwanted attention of ET beings a generation later. Because of her own experience she had feared that the ETs might return to take him away.

Juan looks me squarely in the face and says, simply, *"It was hard for me in the past, but the film, and working with Alan, changed my life. Alan, Jacques Vallee, Nestor Berlanda have dialogued with me and they believe me. I count them as great friends. So though in the past people have called me crazy, today it doesn't bother me. If people laugh at me I send them my best wishes! I am living my life."*

And you can see it in his face. An amazing healing has taken place – a healing that has come through being listened to with respect, and helped to interpret the layers of his experience and the implications of them. Most importantly, Alan Stivelman's work has brought Juan out of his self-imposed isolation into a place of greater joy and community.

Often when people want to be dismissive of those who come forward with reports of close encounters, they will put the experiencers down as people looking for their fifteen minutes of fame. The story of Juan Perez powerfully illustrates how fallacious that idea really is. Film-maker Alan nods as I put this point to him.

"My first sight of Juan was of a traumatised person," he says. *"Juan felt that his case was more like a curse than a gift. As you say, he wasn't looking for fifteen minutes of fame. He doesn't use a cell phone. He isn't used to being on the Internet. He doesn't watch movies. He is a man, living a simple rural life. He is a farmer. A gaucho. When I went to Acevedo he was working on a very big old farm – a hacienda.*

He was a big man – very gaucho as you can see!"

Alan is quite right. Juan is a very big man physically and, notwithstanding his warmth and gentle manner as we talk together, he cuts an imposing figure.

"Juan was trying to explain to me what he saw, what he experienced. I remember that I was filming, recording all the things he was telling me. But then he started to weep. It was not possible to go on filming. So I stopped the camera and I just started to listen."

This, I can relate to. Since publishing *Escaping from Eden*, I have been contacted weekly, sometimes almost daily, by people who have found themselves isolated by anomalous experiences and close encounters. Many tell me of the pain of that isolation and the difficulty of coming to terms with experiences that have broken open their old beliefs and worldviews. Some have found themselves ostracized from their families because their families have been unable to process whatever the experience has been. More than once grown men have wept on the phone as they have shared a story they have kept bottled up for years. People contact me in this way because after all their silence they need someone to do for them what Alan did for Juan, and simply listen. By now I have lost count of the number of people – men in particular – who have reached out to me with stories of close encounters which they have kept to themselves for decades. When they share their story with me, they will often tell me:

"I have told my wife, and I have talked with the person who was with me during the encounter. Beyond that I haven't told another living, breathing soul in the fifty years since."

Our culture is so ready to ridicule and abuse people whose stories don't fit that a great many experiencers choose never to mention what happened to them. For people of faith the mores and beliefs of churches and other faith communities can bring additional pressures. Through my experience in Christian ministry as a theological educator and church doctor, I got to see how Christian faith communities have a way of making it known to their members that their paranormal experiences will be entertained just as long as they can be presented as stories of God, the Devil, angels, demons, humans, animal, vegetable, or mineral. But if you can't tell your story that way, then don't tell it.

When people call me, it is because they need to tell someone – because after years of not talking about it, they still need to process what happened to them. The rate at which this is happening makes me wonder how much more we would hear from our families, friends, neighbours and colleagues if it were not for the ridicule factor. Absent of that threat, I am beginning to wonder if there would be a family or friendship circle anywhere that did not have at least one experiencer or contactee in its number.

The story of Juan Perez illustrates just how isolating it can be to experience phenomena which offend the general worldview. It takes unusual courage, for instance, for an Australian like Jane Pooley to submit herself to public scrutiny on national television simply for the sake of acknowledging her experience. When Ita Buttrose asked Jane to explain why she had left it so long before coming forward with her story Jane said that she had chosen to keep the matter private until after her children had become adults and were more ready to cope with their mother sharing a story that most members of the public would consider ludicrous. Even for the most courageous among us the expectation of ridicule is a powerful deterrent.

Back in 1966 the children who witnessed the Westall Incident were threatened in order to keep them silent as to what they had seen. Fifty years later the fear of ridicule is probably the reason that at least half of the two hundred students concerned are still unwilling to speak about an incident that is in the public domain, even if still unacknowledged by government.

Ridicule, scoffing and shaming keep a tight lid on contemporary stories of close encounters and abductions. Such topics seldom find their way into mainstream conversation, or mainstream media, other than for some light humour. However, our contemporary mores have played no role in shaping the content of our world mythologies and ancestral narratives. These remain from times long ago. And they tell a different story.

By the time I heard Jane Pooley in conversation with Ita Buttrose, I had already made my own mythological journey around the world, from Kenya to the southernmost tip of South Africa, all up the western seaboard of the continent of Africa, into Haiti, Cuba, the Caribbean and as far east as the Philippines. This voyage of discovery has acquainted me with a narrative tradition that occupies a longstanding place in our cultural memory. In that way I can say that generations of ancestors prepared me for a story like Jane Pooley's.

As I listen to her testimony, I wonder how far our culture has really shifted in terms of being able to lift the lid on contemporary reports of abduction phenomena. How close is Western culture today to being able to listen without prejudice? And what might we remember as a society if we can muster the courage to lift that lid and permit others to speak?

Thirty years ago, Australia's television news magazine – *60 Minutes* – took up the question of close encounters and abduction reports, specifically among US defence personnel. To investigate the matter, the show dispatched its award-winning reporter Richard Carleton to interview Professor John Mack.

John Mack was the head of Harvard's Department of Clinical Psychology and a winner of the prestigious Pulitzer Prize for literature. He had recently completed a research project, commissioned by US Defense. His brief was to investigate cases of naval and air force personnel who had filed written reports of close encounters with ETs or ET craft – including reports of abduction. As a Professor of Psychiatry he brought truly world-class expertise to the task of interviewing the experiencers and assessing their psychology and reliability. Essentially, the US Defense Chiefs wanted to know, *"Are these guys sane? And are they safe to fly?"*

Professor Mack began by interviewing more than forty defence personnel before widening his study sample to include civil aviation personnel who had brought similar reports. The conclusions that Professor Mack returned to US Defense were that those he had interviewed were rational and showed no evidence of psychosis. Furthermore, his interview technique had been able to evince secondary details from his subjects which repeated from case to case, thus providing a strong evidential basis for the reliability of their descriptions. In his report for US Defense Professor Mack's recommendation was that whatever it was his subjects were encountering, it was certainly an objective reality which merited further investigation.

Unfortunately, this was a conclusion that neither the Department of Defense nor Harvard wished to accept. Indeed, the report's findings were so embarrassing to both that the professor soon learned that the Board at Harvard was agitating to remove him from his tenure. A committee was established, which proceeded on a fourteen-month review of Professor Mack's methods. As news of this inquiry leaked out, piles of letters from other respected psychologists, psychiatrists and academics began mounting in the Board's mailbox in support of Professor Mack. Even with such stalwart supporters, John Mack's position was under considerable pressure. Ultimately, it was the phenomenal clarity and clout of constitutional lawyer and civil rights attorney, Danny Sheehan, that won the day and ensured that Harvard maintain the professor's rightful position on the faculty. Once Daniel Sheehan was on the scene, the Board rapidly backed down and issued a statement reaffirming *"... Dr. Mack's academic freedom to study what he wishes and to state his opinions without impediment."* The statement concluded with the words, *"Dr. Mack remains a member in good standing of the*

Harvard Faculty of Medicine."

So, to put it mildly, the research brief from the Department of Defense had proven itself to be a poisoned chalice. Nevertheless, Professor Mack, with amazing equanimity, was able to maintain his calm and boldly go where the data led him. Speaking to the *60 Minutes* reporter about his research Professor Mack said,

"After I had worked with forty to fifty of these individuals, I discovered to my amazement that there simply was no psychiatric explanation to this – that something real had happened to them... What has happened to these people is what they say has happened to them. Even though I fully recognize that that's not possible in the worldview in which they and I were raised... There is something profoundly important here that is authentic and real. Either we shrink the phenomenon, or we're going to have to stretch our notions of the possible."

How seriously do we have to take this phenomenon? That was the question. US Defense had taken the phenomenon seriously enough to invite the head of Harvard's Department of Clinical Psychology to assess the situation for them. Professor Mack in turn had taken the phenomenon seriously enough to undertake the research, widen his research base and then publish his findings, knowing it would cost him personally to do so. Yet because Professor Mack had concluded that there was something objective in these cases, the *60 Minutes* reporter, Richard Carleton, felt free to dismiss the Professor's research as, *"So silly, that it's funny,"* and to write off John Mack himself as, *"Off with the pixies!"* The forty to fifty personnel who had been the subject of Professor Mack's case studies were, in the words of Richard Carleton, *"Probably... crazy, off the planet, out of their tree!"*

Clearly confident of his own expertise on the subject, Richard Carleton berated the professor with the words, *"Harvard Professor or not, anyone who believes what you do would believe anything!"*

Precisely what Richard Carleton's qualifications were in the field of clinical psychology I have no idea.

One of the gentlest, most honourable and rational men you could ever wish to know, Professor Mack was a senior academic at the absolute zenith of the field. Yet thirty years ago his carefully researched conclusion regarding the reliability of his subjects was enough, in itself, to entitle an Australian television reporter to dismiss Professor Mack's credentials, rubbish his research and expertise, and count them as nothing.

So, given that context, you will understand why I was open-mouthed to see the chair of the Australian ABC give such a respectful hearing to a woman claiming not only ET contact, but abduction – and not only abduction but co-option into a program of ET hybridization. It was not the only indication I had picked up by 2018 that our culture might be shifting to a place where we may be more ready to stretch our idea of the possible.

In the past when individuals, such as Jane Pooley, have reported ET contact, the mainstream response has been to dismiss and ridicule. When whole crowds have experienced what appears to be ET contact, such as the hundreds at Westall, the official response has been silence. When credible academic voices, such as John Mack's, have endorsed the testimonies of others, we have still felt free to deflect with an insult. But now, in the space of twelve months, two US government departments have come forward to the press to acknowledge publicly their work of engaging with ET technology. Might this be an appropriate moment for us to sit up and listen?

Chapter Four

The Persistence of Memory

Ngambri Country – Today

"Why did you have to bring aliens into it?!"

If you can imagine the tone of a china-shop owner asking me why I let the bull inside, that's the tone.

"I mean I am with you in ninety per cent of what you're saying. I just don't see why you needed to bring aliens into it."

The *"it"* that my friend Adam doesn't want aliens *"brought into"* is his theology. Adam and I trained together back in the eighties and have made much of our journey in ministry in parallel. We have both given decades to the world of church-life, each helping in different ways to keep the ecclesiastical show on the road. While I was planting new church communities, Adam was working to refresh established congregations. When I was training pastors in the interpretation of texts – the Bible in particular – Adam was wrestling with church politics. Adam was exploring the frontiers of deconstructed, organic church while I was serving the Anglican Church as an Archdeacon, troubleshooting communities in transition or in deep difficulty. We have both earned our stripes in the world of ministry and have developed a strong mutual respect.

So when Adam says, *"Paul, let's stay friends, but I won't be reading your book,"* I can't help but feel a bit disappointed. Adam, in turn, is disappointed that I have *"brought aliens into it."* But from my perspective it really isn't my fault. The aliens were there long before I ever noticed.

When an ultimate frisbee injury offered me the opportunity of a study retreat in the shipping crate cabin at the end of our driveway, I knew I wanted to use the time to drill down into some of the anomalies in the book of Genesis. I also knew that there might be an alien or two, lurking somewhere in the hidden layers of its language. Yet I was somewhat unprepared for the full extent of what was to emerge as I began unwrapping the translation questions of Genesis.

"Look, I can see all the same problems you see," Adam explains. *"Morally there's no way you can take the Genesis stories at face value. That's just obvious. Any honest reader can see that what Genesis claims about God is totally at odds with any kind of morality. I don't have a problem seeing all that. But the way I see it is that these are the writings of ancient peoples trying to tease out where God was in the puzzle. There's no need to bring aliens in just to make sense of it."*

For me the red-pill moment came when I realised that the word *elohim* – a word often translated as *God* in the Biblical texts – was in all probability far better translated as *The Powerful Ones*. This is not a small change. In fact, as I quickly learned, making the translation switch alters the whole place of the Bible in the panoply of world mythologies. With a cup of tea in one hand and a Bible in the other, this was the case I now make to Adam one afternoon at our local tea house.

"You see, in the conventional translations, the translators make a call as to when to translate 'Elohim' as God and when to translate it as god or gods, false god or false gods, demon or demons, angel or angels or local chieftain or chieftains. You might want to know how they make that call. With the choices as they are the Bible looks like it contradicts almost every other ancestral narrative of beginnings. It's as if the Bible is saying, 'Forget what all these other cultures have to say. This is what really happened!'"

"But the moment you translate Elohim as 'The Powerful Ones', the Bible stories do an instant about-turn and reveal themselves for what they really are – which is a summary of the stories from out of ancient Sumeria, Babylonia, Assyria and Akkadia. And those original versions are not stories about God. They're stories about the 'Sky People' – another species, who arrived from somewhere else, modified our ancestors, ruled over them and colonised the planet."

"When the Bible was monotheized, in that same moment the history of our ancestors' intersection with Sky People was buried. Yet it's all still there just a translation away from being blindingly obvious!"

Adam silently eyes his fresh cup of tea and stirs it for what feels like several minutes. It's really a brain-twister of an exercise to absorb the implications of these choices in translation. So many beliefs and assumptions are built on top of one another, that there's no real way of fast-tracking all the necessary deconstruction and reconstruction of worldview and theology. Suffice it to say that by the end of our conversation Adam is no closer than he was before to wanting to read his old friend's new book and welcome aliens into his familiar world. Clearly a few more cups of tea will be needed.

I can't be critical of my friend though. It is a total reframing. But it happens because of what's already in the text. The translation key I mentioned to Adam not only reveals the family connection of the Bible and the Sumerian texts but it affirms themes in a great many of the stories of beginnings curated by indigenous cultures all around the world.

That's why today I am looking out over the Australian bush of my hometown with revered Aboriginal Elder, Shane Mortimer. Shane is an incredible person – a man who simultaneously embodies charisma and calm, peace and passion. He is a powerful advocate for his people and for the land of Australia. Early in the twentieth century, the site for the new Australian Capital Territory was identified by the colonial government, and the authorities began rounding up the local indigenous people to remove them from the land and make way for a new, white capital city. Shane's family was one of the first to be removed from their ancestral lands. The only reason that Shane's aunties and uncles did not become part of the *"Stolen Generation"* was that his grandfather was wealthy enough to be able to bribe the Catholic nuns with sums of money large enough to persuade them not to forcibly take his children, to be raised in one of the country's child detention centres, euphemistically referred to as *"missions"*.

The purpose of Australia's *"Stolen Generation"* policy was to extinguish Aboriginal culture – its ways of life, its values, its language and all the story it carries. Story, however, has its own way of surviving and resurfacing. Elders past and present carry their people's story and pass it on orally from generation to generation. So, it is with some anticipation that I prepare to sit at Shane's feet and listen.

Shane would have every right to carry some deep resentments, and to feel ill at ease among the European latecomers to his country. As an economic migrant from Europe to his ancestral home I might have felt a bit sheepish in the presence of one of our country's Aboriginal owners. Instead Shane welcomes me warmly and we walk together along the grassy slopes of Nadya. As I listen to Shane, I am learning stories that have been carried by the original Australians for some sixty thousand years. Among these stories are mysterious Dreamtime narratives which speak of who we all are and where we all come from.

Late in C18th the early British colonisers of New South Wales noted their Aboriginal neighbours' strong attachment to the land. They knew their land, how to read it, how to preserve it and how to farm it. It therefore came as some surprise that when asked, *"Where do your people come from?"* Aboriginal elders pointed the newcomers not to Australia's red ochre soil but to the stars. And not just any stars. Their ancestral source, according to this narrative, was a world orbiting a star somewhere in the constellation we call the Pleiades.

This is not the first time I have heard of such a connection with the Pleiades. At a gathering of researchers in Sydney, Australia I got deep into conversation with a pastoralist called Blair. Blair's family farm (or *"station"*, as we call it in Australia) was home to hundreds of sheep and cattle. It was also a place that seemed to attract UFOs. Indeed, on that station groups of craft were witnessed regularly, and by groups of people, over a twenty-year period. When he told me this I had to ask,

"Blair, do you think there was something about that place or about your family that made your visitors feel free to keep turning up? I mean wasn't your family afraid?"

"Oh no," said Blair. *"My family are Cherokee. My father taught me that some of those who come have been visiting us since the beginning of our people's history. They came to our ancestors and taught us how to live as people on this planet. Taught us about foods and medicines. And they told us where they were from – a planet, orbiting a star in the Pleiades."*

The great leap forward that our ancestors made, from living on the planet in animal subsistence to the civilized life of a crop-farming, city-building civilization, is an anomalous moment in human prehistory that has long fascinated paleobiologists and which begs for an explanation. The story of our being assisted in this great leap by prehistoric visitors from the Pleiades is an indigenous explanation, handed down through the ages from one generation to the next by Aboriginal Australians and by Cherokee people alike.

Today as we stand together at Nadya (the slopes of Mount Ainslie) Shane shows me land that has been home to such stories for tens of thousands of years. We walk up to a place of corroboree. Its level floor is a shingle of tiny stones and even tinier fragments of animal bone. The floor is bounded by the natural amphitheatre of a ridge of rock, behind which is a magnificent view of the valley, backgrounded by brown hills and the blue-green mountains on the far horizon. For millennia Aboriginal men gathered here from tribal communities scattered along a rocky ridge in the continent, a thousand miles long. The amphitheatre of sitting rocks is part of that ridge. On these same rocks men have sat and conferred, renewing friendships and accords as they swapped the sharp, bone tips on their hunting spears.

"Where we are standing now is where the elders prepared the young men for initiation. See these stone markers? See how they have been carved to make a viewfinder. It points across the valley to the top of Tidbinbilla. That is where the initiation would happen. The elders would show the boys the plain below, what to look for; which fires meant invasion, which fires meant a request to come on to country."

It is a beautiful, natural vantage point. From here the elders could show the boys what was the men's work in protecting their tribe and their country – what their place was in the community.

"Behind us is the mountain for secret women's business. That's not something I can tell you about!"

This is an incredible place. The cultural significance of Nadya would be difficult to exaggerate. I feel privileged to have my feet on this soil. At the heart of the site is the beautiful petroglyph of a kangaroo, carved with great beauty and elegance into the granite. It is a work of art that would have taken many years to create – an enormous effort simply to carry one aspect of this great canon of storytelling. I breathe in the magnificence of this sacred place. There's no doubt, this should be a world heritage site.

"It should be," says Shane. *"But it's been sold to a developer for a complex of townhouses. Where you are standing is designated for a tennis court."*

This is only one of the latest assaults on the world's oldest continuous culture. Indeed you can draw a straight line from that moment fifteen minutes after Captain Cook's landing at Botany when crew of *HMS Endeavour* began shooting the local Aboriginal people, to the events of 2020 with the blowing up of Aboriginal caves in the Juukan Gorge of Western Australia and the would-be destruction of this sacred site in Canberra.

When the mining corporation Rio Tinto went into the Juukan Gorge they found a 4,000 year-old genetic link to present-day traditional owners along with evidence of continual human occupation going back at least 46,000 years – making it the only inland site in Australia demonstrating continual occupation through the last ice age. The history carried by the place is incredible and unique. But this year Rio Tinto blew it up with not much more than an anaemic apology for any *"offence"* caused.

"They call us the 'traditional owners'," says Shane. *"And what that phrase does is convey the message that our people are the former owners. Now this land belongs to someone else!"*

"They call my people 'prehistoric'," he says. *"It's a way of saying that anything Aboriginal Australians have to say about the past is outside of real 'history'. Only White story is called 'history'. What Aboriginal Australians say doesn't even count as history!"*

This is the grim way conquests go. When new Empires arrive in a country or whenever new regimes establish themselves, they have to take charge of history. It is the colonisers who will shape the syllabus in education, and the history celebrated in statues and carvings. All other truths must be extinguished.

That is why Aboriginal Australian narratives of humanity being seeded from the Pleiades is a story absent from our school curriculums. The colonisers didn't endorse it. It is why the Native American stories of star people arriving and nurturing our ancestors' abilities in nutrition, medicine and farming also have also not made it into school textbooks and official histories. The Europeans didn't originate the story therefore it is not history, it is *"mythology"*. And for *"mythology"* read *"fiction"*.

This is the way we are taught to ignore the body of history curated by indigenous peoples. Because it will survive orally, and for that reason cannot be completely extinguished without extinguishing the culture itself, indigenous story must therefore be framed as folklore or fiction. That way it can never address questions that we consider historical or scientific. The folklore/fiction label teaches us to expect only cute stories or irrelevance from the world's ancestral narratives.

The removal of ET narratives from our official canons of science and history is a pattern that recurs through the ages. I first saw this pattern as I sifted my way through the book of Genesis, tucked away in my shipping crate. Reading the ancient text with *elohim* as a plural surfaced a whole new reality – a forgotten world in which many different kinds of entities cohabited and interacted with our *"prehistoric"* ancestors – a forgotten world in which a body of ET colonisers duke it out over the ins and outs of project Earth.

Today, there is a broad consensus among Biblical scholars that this mishmash of *elohim* stories received a drastic edit sometime in the sixth century BCE. By this time Judaism had become firmly monotheistic. This meant that the memory of other *"gods"* – the echoes of the Sumerian stories of Sky People – all had to be airbrushed out of the Hebrew Canon. The problem of too many *"gods"* could be dealt with – so the editors must have thought – by translating the word *"elohim"* as *"false god"* or *"demon"* in the texts where the *elohim* concerned were clearly on the wrong side, and then translating it as *"God"* where the *elohim* concerned appeared to be in charge. Problem solved!

Except that the previous shape of the stories remains. Clues surface whenever *"God"* does brutal and inhuman things, or when *"God"* opposes human progress, or fails to anticipate events that even a child could see coming. Yet the translators clearly believed they were being faithful to the Hebrew tradition by cleaning up the texts this way and layering new meanings and imperatives over the top of them.

The editors may have reasoned that Moses himself had commanded that these other *elohim* be forgotten. For instance, the Ten Commandments famously begins with the commands,

You shall have no other "elohim" before me. You must not bow down to them or even depict them.
(Exodus 20:3-5)

It appears that a great forgetting is being commanded.

Moses' successor Joshua effectively repeats the same command as he inherits the mantle of leadership. He says,

> This is what YHWH the Powerful One of Israel says: "Long ago your ancestors, including Terah the father of Abraham and Nahor, lived beyond the [Euphrates] river and served other [powerful ones]... Now fear YHWH and serve him with all faithfulness. Cut yourselves off from the [powerful ones] your ancestors worshipped beyond the [Euphrates] river and in Egypt and serve YHWH... Choose for yourselves this day whom you will serve. Will it be the [powerful ones] your ancestors served beyond the [Euphrates] river, or the Powerful Ones of the Amorites, in whose land you are living?" As for me and my household, we will serve YHWH.
> (Joshua 24)

It was only when I started reading the *elohim* as plural entities – indeed as another name for the Sumerian's Sky People – that it became clear to me that a policy of censure and forgetting is actually embedded within the story of the Bible itself. And the final edit of the Bible does not mark the end of the forgetting. As I revisited the beginnings of Christian history my studies led me to significant Church Fathers who leaned away from accepting the Hebrew Canon at face value, and leaned towards a quite different story of our emergence, one summarized in the works of the Greek philosopher Plato.

These rather interesting Church Fathers included people like Clement of Alexandria, Justin Martyr, Origen and Marcion. All were foundational leaders within the kaleidoscope of early Christianity. Each in their turn argued in effect for an Old Testament of Plato and a New Testament of Jesus Christ. They regarded the *elohim* narratives of the Hebrew scriptures as being something other than stories about God, and interpreted them in a way that harmonised with Plato's explanation of beginnings

– an explanation which sounds many of the same notes as the Aboriginal and Native American stories I mentioned before.

However, this broader view was soon to be shut down by the institutional gatekeepers. In 144 CE Marcion was excommunicated. The C6th BCE edit of the Hebrew Scriptures was ruled in and Marcion's Platonised interpretations were ruled out. In 543 CE Origen was condemned as a heretic, having powerfully influenced Christian theology for three and a half centuries. His was the most influential voice of those times subtly directing God-seekers away from the edited Hebrew stories and towards the writings of Plato.

By those actions the gatekeepers of the Church institutional sent out an emphatic signal – namely that Platonic ideas – which included the stories of a big universe, populated with an array of entities, some of whom had interacted with and adapted our ancestors – were to be understood as being totally out of bounds. Orthodoxy would have no room for them.

The removal from Christianity of these broader views was really concretized by the C4th CE Roman Emperor, Theodosius. In 381 CE, to help all good citizens be good Christians, Emperor Theodosius decided it would be expedient for him to wade into a theological debate, then raging in the Church, and settle it with a stroke of the imperial pen. Theodosius' legislative input had a number of, probably intended, consequences.

Firstly, it cemented the ranks of hierarchy in the institutional church as a pyramid of powers with the people at the bottom, the Bishops sandwiched in the middle, and the Emperor emphatically at the top.

Secondly, it settled decades of public conversations and conflicts in which violence and murder were often the tools of theological debate. Theodosius' legislation imposed an official peace within an institution that he considered ought to harmonise rather than polarise the good citizens of the Empire.

Thirdly, and most importantly of all, Theodosius'

promulgation clarified that it was the Emperor who was the ultimate arbiter of what was true and what was false. Any ruling power or conquering force has to take charge of the narrative of history. It was the job of the Empire to provide the people with its *"Department of Information"*. This was not something that could be left to local chieftains, indigenous priesthoods or ancestral traditions.

The era of *"Kaleidoscope Christianity"* was emphatically over. Orthodoxy had landed.

On the face of it, Theodosius was guillotining a longstanding tussle over a fine point in Christian theology. But what he had really done by stepping in so decisively was to militarise the forces of orthodoxy and turn into rebels and subversives all the leaders of the now officially *"fringe"* spiritual and religious sects. Henceforward every indigenous priesthood with a story of its own was to be considered suspect and might be better disbanded. Every written text proposing narratives of human origins and other-worldly contacts at odds with Orthodoxy's neat and tidy creation story would need to be removed from the shelves of acceptable literature. They were now texts to be archived or burned. Not taught.

Uniformity was the order of the day. The vast array of languages and alphabets, scattered throughout Rome's international territories, were pushed down to the grassroots. Business, law, science, education and religion were all to be conducted in the Imperial language of Latin. Suppressing local languages and ancient alphabets served many purposes, not least that it buried the histories and ideas expressed in the old languages and etched in foreign carvings and glyphs. It was not the last time an empire would attempt to silence ancestral story.

In the fifteenth century, the Catholic imperial forces of Portugal and Spain began their invasions of Central and South America to claim their respective pieces of the New World. As they established their hegemony over the indigenous peoples of the land, they did the exact same thing Emperor Theodosius had done more than a thousand years before. Every agency of control that had followed piecemeal in the wake of Theodosius' promulgation was now combined into a swift and decisive methodology, which the invading forces rolled out in rapid succession, to quickly establish full spectrum dominance over their new territories.

All indigenous priesthoods were rounded up and summarily executed. Their books and scrolls were confiscated and burned with, perhaps, the best specimens sent to Rome to join the vast library of texts curated by successive generations of emperors and popes.

Among the suppressed South American texts were those of the Mayan tradition. Its account of human origins differed significantly from the version of events now being promulgated through the Catholic schools and churches which now peppered the landscapes of South America. It was the job of these institutions to explain definitively to every person resident in the newly Christian territories the official explanation of where we all came from.

The Catholic authorities may have hoped that the countries' indigenous history was well and truly dead and buried, being either under lock and key or in cinders. However, time would tell that Mayan memory was far from lost. In the next chapter I will take you to the mountains of Guatemala, where we will discover that the Mayan tradition, like all ancestral memory, has a surprising longevity and a curious way of returning from the dead.

Chapter Five

Voices from the Grave

Guatemala – 1703

We are in the highlands of Guatemala, where a young Dominican Friar by the name of Francisco Ximenez has arrived from Spain to serve as a parish priest in the mountainous district of Chichicastenango. Francisco is a true enthusiast for the Dominican tradition he represents. This means he has a passion for scholarship and the study of ancient books. He is an avid linguist and relishes the opportunity to sit and listen to the stories of his new hosts as he learns the ins and outs of the local language.

Francisco's great enthusiasm, his warmth and his genuine respect for his hosts gradually wins the trust of the local community, who are the successors of the Mayan people, whose culture once dominated this country. And then one day Francisco is presented with an unfamiliar text, written in the indigenous language of K'iche'. The *"Priests of the Feathered Serpent"* – descendants of an ancient Mayan priesthood – explain to the young Francisco that this ancient K'iche' text holds their story of human origins. Somehow this text has survived the banning, burning and burying of the fifteenth century. It is a story which has been told for at least one hundred generations, detailing the Mayan succession of kings and explaining how the human race itself first came into being.

Seven years go by as Francisco navigates the nearly forgotten K'iche' language and finally emerges with a translation into eighteenth century Spanish. What has surfaced through his literary journey is an account of human origins, kept hidden for more than ten generations. Francisco Ximenez's translation becomes known as the *"people's book"* or *"Popol Vuh"*.

The *Popol Vuh* tells us that in the beginning the Earth was flooded and shrouded in darkness. Cloaked in the blackness of the dark sky powerful beings arrive, whom the writer describes as *"Those who Engineer."* As the enigmatic Engineers hover over the waters of the dark and flooded terrain, they discuss how they might nurture life on Earth; life in the waters, plant life, animal life and sentient life.

Let's just pause there a moment. What are we being told? Are the mysterious Engineers flying, individually, bird-like, as they assess their new territory and take counsel? Is this how they have arrived from interstellar space? Or are they hovering in something?

This is a moment in which many world mythologies reinforce and finesse one another. The first time I read an English interpretation of the work of Francisco Ximenez I was mind-blown by the parallels I found. To name just five sources, the Elohim narratives of Genesis, the Sky People narratives of the Sumerians, Filipino mythology, the Osanobua narrative of the Edo people of Southern Nigeria and Benin, and the description of the *Popol Vuh*; all begin with the waters of a dark and flooded planet. In the Hebrew text, the planet is *"tohu wa bohu"* – a phrase which means devasted and chaotic. It as if the planet has been laid waste by some cosmic event. And it exists before the work of *"creation"* which reboots the planet and nurtures the forms of life Earth with which we are all familiar.

The first thing that has to happen for this planetary rehabilitation to begin is the clearing of the skies to allow the sun to begin driving all its forces of life. Land is then reclaimed and the waters are separated into salt water for the seas and freshwater for the land, and the land has to be cleared. The flooding has clearly been at an oceanic level. Only when this has been done can the zoological work begin.

The *Popol Vuh* tells us about the Engineers arriving from somewhere and hovering over the waters, high in the dark sky, discussing how this is to be done.

Similarly, the Edo account opens on a planet that has been completely flooded. The sons of Osanobua then arrive and begin reclaiming land. Osanobua himself descends from out of the sky on a long chain, stretching as far as the eye can see into space. He is *"The Almighty One Above the Waters"*. Once stationed above the waters he delegates to his sons the work of terraforming the planet and managing project Earth.

The Filipino narrative is of the arrival of Tagalog, a giant bird, who hovers hawk-like over the waters. Tagalog then creates vortices of wind which pull the waters away from the higher ground to create the islands.

Similarly, the Sumerian narrative begins with the descent of the Sky People from the heavens. They too create a number of strong winds which draw the waters away from the land and then separate the waters into oceans of salt water and freshwater rivers and currents.

Genesis tells us that the *Powerful Ones* arrive over the dark floodwaters in a *"ruach"*. The word ruach is popularly translated as the *"Spirit of God"*. The text specifies that the *ruach* was *"hovering"*. This word for hovering – *"merahephet"* – is the word the Bible uses to describe how birds of prey hover in the sky, appearing to float in the air, without moving their wings. This is what the *Powerful Ones' ruach* was doing. It was hovering in the sky without moving any wings.

I learned this detail from Mauro Biglino – a scholar of ancient Hebrew, who for many years worked for the Saint Paul Press in Rome, translating with great precision the literal meaning of Hebrew words for Vatican-approved interlinear Bibles. Providing the interlinear meaning is a very exact discipline. The scholar must be rigorous in avoiding any kind of interpretation of the word, and give only the literal, etymological meaning of each word part. So it is with that degree of precision that Mauro explains in his writings that the use of the *ruach* elsewhere in

Hebrew literature shows that the word means either a *"wind"* or *"something flying through the air and creating a wind"*.

Interestingly, this meaning has survived in the Ethiopian (Amharic) word *"roha"*. It means a fan or anything that moves through the air and creates a breeze.

What could this wind-making *ruach* possibly be? Mauro Biglino argues that *ruach* may be a loan word from the ancient Sumerian language. He notes that the pictograph for *ruach* in archaic Sumerian cuneiform depicts two elements. The first element is a body of water. Hovering above the water is the second element – a motif which could be seen as a giant eye – although to the C21st viewer it is difficult not to see it as a flying saucer. A flying saucer hovering over a body of water is pronounced *"ru-ach"*. Get the picture?

Now it's fair to say that not every expert in the field agrees with this reading. But to corroborate Biglino's interpretation the Prophet Ezekiel gives us a significant help in a later Biblical text. In his book Ezekiel takes quite some time to describe in great detail what a *ruach* actually looks like. The one he describes is on omnidirectional wheels and is metallic and silver in colour, with a canopy of crystal-clear glass. The particular *ruach* he saw and flew in arrived in Iraq from the sky. It flies, transports people, and makes a loud roaring sound when it moves. There is no question, we are being shown a craft.

The incredible correlations in ancient accounts from Nigeria and Benin to Iraq, from the Philippines to Guatemala, confirm that what we are being shown is a visual memory. Each tradition has found its own language and metaphor to verbalise the same kind of eyewitness account. Ancestral memory from all around the world has sought a way to tell us how something looked, to let us know what it was our distant ancestors saw when the Powerful Ones/Sky People/Rulers from the Sky/Engineers first arrived on a flooded planet Earth, shortly after our planet's last great cataclysm.

Together this array of ancestral narrators overturn our familiar story of beginnings and resolve the questions of interpretation raised for us by Genesis and the *Popol Vuh*. The Powerful Ones flew here in a craft, which hovered over the flood waters and began terraforming a devastated planet.

These global parallels make clear for us in the C21st what Francisco Ximenez was able only to guess at as he allowed the unfamiliar themes of the K'iche texts to take shape in C18th Spanish. Yet what did take shape was more than clear enough for Francisco to easily understand why this indigenous Guatemalan story of beginnings had seemed so unacceptable to the imperial forces that had banished and buried it two centuries before.

Following their dramatic arrival in the *Popol Vuh*, the Engineers then commence a sequence of experiments aimed at producing a species of beings who will be intelligent enough to serve as a useful workforce but with sufficient inbuilt limitations to make them easily managed. Just the same as in Genesis, the figure responsible for upgrading the humans from an animal state is identified with the serpent. Q'uq'umatz aka Kukulkan aka Quetzalcoatl is the flying or *"Feathered serpent"*.

As the story goes, Quetzalcoatl's first attempts to engineer a useful slave species produced unsatisfactory results. The resultant creatures were either not intelligent enough to serve, or they had no interest in *"worshiping"* or serving their superiors. So they were eradicated with a flood. The survivors of these unsuccessful genetic experiments are, according to the *Popol Vuh*, the ancestors of the ape-like creatures who live in the forests. (A rather intriguing footnote, which has us and apes sharing a common primate ancestor.)

Quetzalcoatl's next attempt produced a magnificent result. Homo Sapiens Quetzalcoatlus was everything that we are and more. In addition to having all our cognitive abilities, Homo Sapiens Quetzalcoatlus were able see beyond what was presented to their immediate senses. They could anticipate events before those events unfolded. Consequently, they were not easily deceived. They were clever, and the hint is they could even remote view. This, of course, left their rulers without much of an advantage.

Feeling rather threatened, the Engineers confer and quickly agree that Homo Sapiens Quetzalcoatlus is clearly too advanced – far too conscious to be easily managed. So the Engineers then set to work again. This time their research leads them to manufacture a special vapour. When sprayed over human populations the vapour debilitates the humans' neurological function, diminishing their abilities to a more manageable level. Thus adapted, Homo Sapiens Sapiens were dumbed down sufficiently to see only what they were shown and understand only what they were told – just intelligent enough *"to be avatars for us (The Engineers) to work for us and bring us our food."* We were the satisfactory result.

This was the story of our origins which the Roman Catholic colonisers had done their best to extinguish back in the late fifteenth and early sixteenth centuries. The indigenous story was clearly an affront to the neat and tidy narrative of Christian orthodoxy. Yet despite the colonisers' brutal attempts at killing the old, old stories, the indigenous folklore had managed to survive and resurface, keeping alive for posterity the Mayan memory of human beginnings.

Back in Australia I find myself again on the slopes of Nadya, surveying the place of corroboree, the mysterious petroglyph, and the young men's initiation site. To know that this sacred place in my home city has been zoned for townhouses and tennis courts is to know that the ages-old pattern of colonisers obliterating indigenous knowledge is a story as alive as ever. Over the last twenty years we have all seen it on our screens as the forces of *"ISIS"* have defaced, desecrated, and destroyed some of the most precious and ancient built heritage on planet Earth.

Some of my friends in Defence have seen it first-hand. In the pages of *Escaping from Eden* I tell the story of Jorg Fassbinder's team. His was an archaeological team that entered Iraq with the protection of allied forces within days of George W. Bush's incursion into the country in 2003. Fassbinder spoke excitedly to the BBC explaining that the site they were entering showed evidence of being the tomb of the Sumerian King of Legend – the hybrid being by the name of Gilgamesh. What would they find? What might DNA analysis reveal? Here was a once in a lifetime opportunity to test the claims of the ancient Sumerian narratives of Sky People modifying our ancestors. Were these stories pure fiction? Or were they rooted in fact? Those of us who heard the story held our breath. But within days the story went silent – and in the nearly two decades since the find we are expected to believe that no further investigation has been made.

As *Escaping from Eden* has begun to make its presence felt, I have come to understand – through first-person testimony – that the story of the Gilgamesh find was just one part of a bigger and even more fascinating picture. At the time of the 2003 invasion of Iraq the world looked on, uncertain as to the motive behind the war. Was it simply about access to oil? Was it to insert a new regime more amenable to US interests? Was it to confiscate weapons of mass destruction? Of course, wars have many layers and many agendas may be served by a single incursion. The confusion as to why we are aggressing our way into someone else's country is felt most painfully of all by the brave servicemen and women who have to effect the invasion. It has been from this quarter that I have learned that on arrival in Iraq, among the units deployed some were tasked to manage military threats, and others to quarantine areas of political and economic importance. There were yet other units who were deployed with a more intriguing assignment. Their mission was to obtain artefacts from archaeological sites and remove them to the USA for analysis and protection.

It would be difficult to exaggerate the importance of such retrievals. The archaeological sites in question relate to the most ancient known narratives of human civilization – those of the ancient Mesopotamian cultures of Sumeria, Babylonia, Akkadia and Assyria. Any artefacts from the sites concerned would have a direct bearing on the authority of narratives that predate every civilization and every religion on planet Earth today. What then were the recovered artefacts? What were the stories they carried? And why did they have to be removed and secreted away? Are they being protected for posterity? Or have they been hidden from history?

Back in the C4th CE when Theodosius' legislation off sided any narrative that dared to contradict the new Christian orthodoxy, the many guardians of indigenous history knew they had to safeguard their own artefacts from the *"protection"* of Imperial forces. If their ancient narratives were to survive they would have to do so in texts buried in caves in the Nag Hammadi desert, or in the underground meeting places of esoteric and gnostic sects, the knowledge passed ceremonially from one generation to the next. Sometimes the information would journey, hidden in plain sight, encoded in the metaphors and imagery of esoteric poems and myths. It was from the inspiration of such esoteric sources that later generations of scholars would be made to wonder about our true origins and the possibility of a wider cosmic family.

One such student of esoteric knowledge was another Dominican brother – the Italian friar, Giordano Bruno. Unfortunately, what happened to Giordano Bruno was to stand as a warning to four centuries of Christian believers never to dig up what our rulers have buried, never to learn forbidden languages or probe forbidden texts, and certainly never to bring ETs into the conversation.

Chapter Six

Secrets and Revelations

Campo de' Fiore, Rome, Italy – February 6th, 1600

In space there are countless constellations, suns and planets. We see only the suns because they give light. The planets remain invisible, for they are small and dark. There are also numberless earths circling around their suns. It is unreasonable to suppose that those teeming worlds should not bear similar, or even more perfect inhabitants than our Earth.

These are the words of an inspirational speaker, a friar of the Dominican Order. His ideas about the cosmos are stimulating and stretching but they are not entirely new. Two and half millennia ago Hinduism's Vishnu Purana shared a similar vision. It spoke of planet Earth as being one of billions of inhabited planets, scattered throughout the cosmos. However, what is considered possible in the East is totally unacceptable in the Catholic West, and Giordano Bruno's repetition of this cosmic vision is one of the many things he has said that have brought him to this public square and sealed his grisly fate.

For the last seven years Giordano Bruno has been in prison, where he has been held without trial in the hope that he might recant his heretical ideas. Such a recantation might have broken the spell of his popularity among the public who flocked in great numbers to his lectures and demonstrations of mental skills. A recantation would serve as a warning to Christians throughout Europe never to stray too far from Catholic orthodoxy. But Bruno has not recanted, and so a trial has been hurried through to allow for a public execution today that will send shock waves through Christendom for decades to come and leave people in no doubt that the ideas of Giordano Bruno are ideas to be avoided at every possible cost.

Armed Papal Officers standing on the street corners are marshalling the crowds and encouraging them to cheer as the filthy, bloodied object of a man is dragged into the market square, called the *Field of Flowers*. Few would recognize him as the charismatic scholar who only a few years ago was dazzling Europe with his speaking tours. His public demonstrations of phenomenal mental skills, and his various books on mental self-improvement and mnemonics have made him a popular figure in a Europe hungry for new ideas. Today he is an object of horror.

The Papal guards unchain the man and fix him on a stake, upside-down. One guard now pulls the man's tongue forward out of his mouth and nails it to the stake. Next, he draws a knife from its scabbard and uses it to sever the man's tongue. Other officers of the Church now step forward to ignite the pyre positioned below the man's head. The crowd watches in fearful approval as Giordano Bruno goes up in flames. It is Ash Wednesday.

So, what were the dangerous ideas that ended the life of this brilliant Dominican friar so tragically?

Firstly, Bruno was a man well ahead of his time in embracing the science of Copernicus, which asserted that the Earth orbits the sun, not vice versa. Bruno, however, went even further. He claimed that the stars we see are in fact other suns like our own, and that just as Earth orbits the sun, so other planets orbit those other suns, planets which may be home to other beings, other people, other civilizations. The words I quoted at the top of the chapter were written by Bruno in 1584. Bruno's studies in Plato and the intellectual children of Plato led him to the view that our universe is ever-expanding and that the process of creation is still unfolding, generating space, time, and matter from an infinitesimal zero point. The consciousness we experience is part of the consciousness of our planet and cosmos. Our consciousness precedes our life on Earth and takes us to other realms – maybe to other parts of the universe when our life on Earth concludes. Our consciousness unites us with the Source of the universe. It can be heightened and developed in a way that transforms our potential in this life. This was the belief that had motivated Bruno's hugely popular speaking tours.

Where Giordano Bruno really ruffled some feathers was in his teachings about Jesus Christ. For Bruno, Jesus reveals Divinity to us not because he is different to the rest of us, but because he is the same; a soul on the same great journey as the rest of us, only far more conscious of his identity with the Source. When we look to Jesus we are looking to a model for our own life. It was an exciting and empowering belief, but it bumped up against Catholic Orthodoxy's emphasis on the divinity and uniqueness of Jesus.

Bruno's beliefs about human consciousness left no room for the fear of hellfire with which his opponents tried to threaten him. In fact, when his sentence was pronounced in an ecclesiastical court, he famously looked at his prosecutors and said, *"I can see in your eyes that you are the ones who are afraid. Not I!"*

Given how emphatic the Church's refutation of Giordano Bruno had been I was in a state of open-mouthed disbelief when in 2009 I heard the same institution challenging the world to reconsider the possibility that humanity may be part of a much larger cosmic family. Christian believers ought to be ready, they said, not only to believe in but to embrace a *"brother or sister alien."*

These were the conclusions of the Pontifical Academy of Sciences after a five-day Colloquium of closed sessions in which an exclusive selection of senior scholars and theologians talked through *"The Theological Implications of Contact with Other*

Civilizations".

Reverend Doctor Guy Consolmagno, the Senior Astronomer for the Vatican Observatory at Mount Graham, Arizona, insisted that believers really shouldn't be surprised to encounter intelligent extraterrestrial beings. They are creatures of the same Creator, and children of the same Heavenly Father. We should not be surprised, so he said, because they're in the Bible – both in the Old Testament and the New Testament. This was a challenge, and no mistake; an invitation to spot the alien – like a theological version of *"Where's Wally?"*.

Monsignor Corrado Balducci, a senior advisor to the Pope concerning demonology and the paranormal, added another element to the conversation when he wrote that people who report close encounters or abduction experiences are not reporting something demonic or a psychotic episode of some kind. They are reporting a totally different kind of being – one that merits serious study.

That's a shift. Coming from the spokespeople for Benedict XVI, the most conservative pope in my lifetime, those statements represent a huge shift! What was forbidden in times past was now being placed squarely back on the table by Roman Catholic authorities. Fr Gabriel Funes, the Director of the Vatican Observatory, said on more than one occasion that the faithful should be ready to embrace an ET neighbour, *"Sooner than anyone anticipates."* The message was clear. *"People get ready!"*

But why? Why had they taken this taboo subject off the shelf? Why did Vatican spokespeople challenge the faithful to be ready for a disclosure of contact with another civilization with such a sense of urgency?

It made me wonder. Does the Vatican know more than it is saying? Does it possess information that it is not disclosing? The press statements, articles and interviews following the Colloquium of 2009 represented a 180 degree turn in the Curia's historic position. Back in 1600 Giordano Bruno's public humiliation and excruciating execution was, of course, a death threat towards Bruno's followers and indeed to any scholar who might choose to investigate such forbidden topics. Just to press the point home, within three years of his death by immolation, Bruno's own writings were added to the *"Index Librorum Prohibitorum"* – the Roman Catholic Church's official list of books that Catholic believers are not allowed to read.

Of course the use of death threats to silence discussion did not disappear with the turn of the seventeenth century. In the C20th it was a practice still alive and well in the USA, being employed as a way of excising the discussion of extraterrestrial life from the public sphere. Of course, from time to time, death threats have to be followed through and it was difficult not to be aware of this, growing up where I did.

Great Britain
1982-1990

Every morning and every afternoon I cycle the six miles between home and school, along a busy stretch of road in Buckinghamshire. Each day the journey takes me past an enigmatic building, set back from the road, behind the security of a guard post and boom bar. Everyone in the locality knows that it is a research facility, staffed by a privileged elite of our scientific research community. There is even a sense of local pride in the place because the technicians at this facility have been engaged to help develop state of the art weaponry commissioned by President Reagan for deployment in space. These will be the first off-planet armaments ever developed. In a speech at the United Nations, President Reagan has framed the importance of this top-secret project in dramatic terms:

Perhaps we need some outside universal threat to make us recognize this common bond. I occasionally think how quickly our differences worldwide would vanish if we were facing an alien threat from outside this world. And yet I ask you, is not an alien threat already among us? What could be more alien to the universal aspirations of our peoples than war, and the threat of war?

The Reagan administration will be investing $53 billion into *"Star Wars"* – a military response to the aforementioned *"alien threat"*. Its phenomenal price tag, and the fact that *"Star Wars"* is high technology for warfare off-planet, has raised more than a few eyebrows. My friends and I wonder if the President's cryptic words are about an alien *"threat of war"* or a threat of *"alien war"*. Is that what our *"Star Wars"* defences are really for? If these questions haven't given us reason enough to wonder if something odd might be going on, the suicides of twenty-five scientists working on the project at the mysterious facility down the road certainly have.

As the drama plays out, one by one, the families of the victims come forward to contradict the company's story. No, these deaths were not suicides. The families want to let the world know that their son, their husband, their father, was not distressed or depressed, prior to being found dead. The families completely reject the official story intended to explain away the sequence of deaths at the company.

The various coroners involved are also not playing ball with the official story. One after another the coroners return open verdicts with only a couple of cases strangely classed as *"misadventure"*. Now the victims' union leader steps into the fray, Clive Jenkins, General Secretary of *the Association of Scientific, Technical and Managerial Staffs (ASTMS)*. Categorising these deaths as suicides is, he says, *"statistically incredible."* His association members are deeply concerned by what he calls, *"these clusters of suicides, violent deaths or murders."* Clive Jenkins is calling it for what it is and is insisting on an urgent government enquiry. Margaret Thatcher's administration, however, considers an enquiry to be quite unnecessary since the relevant minister is more than willing to bring us the official explanation. The sad succession of suicides is, he says, no more than a sequence of unfortunate coincidences. Twenty-five of them. There will be no enquiry.

The tragedy of these deaths through the 1980s was a powerful reminder that we were still deep in the era of silence and denial. Evidently, lessons have been learned since the days of the Inquisition. Why silence communities of people with public shows of torture and immolation (such as in the case of Giordano Bruno) when a conspicuous number of unfortunate coincidences can serve the exact same purpose?

One man who deplored the use of death threats against witnesses of UFO phenomena was Apollo astronaut Ed Mitchell – the sixth man to walk on the moon. Ed Mitchell worked tirelessly to disempower these kinds of threats, which he knew to have been put upon his friends and fellow townspeople from Roswell, New Mexico. At the turn of the new millennium this commitment was on full view.

The Washington Press Club – Washington DC
May 9th, 2001

Ed Mitchell is standing at the podium of the Washington Press Club. The room is buzzing with more than sixty military, corporate, government, civil aviation and scientific experts, all present to testify before an eager crowd of photographers, journalists and reporters. Some are ash white. Others are weeping. For some of them, today will be the first time they have breached their security orders to speak publicly on their knowledge and experience of the UFO phenomenon and of an extraterrestrial presence.

Ed Mitchell is speaking:

Some of the old people who had been involved at the time – what I call the "old-timers" – who were involved... in 1947 with the

Roswell crash – people like in the Sherriff's department who had been to the crash site, and were supervising traffic; my friend and the friend of our family, the Major who was in an administrative function at the Walker airbase; these folks, because they had been shushed and told not to talk about their experience, by military authority, and on dire consequences if they did – felt that now (in the 1970s) many, many years later that they did not want to go to the grave with their story. They wanted to tell someone reliable. And being a local boy, and having been to the moon, they considered me reliable enough to whisper in my ear their particular story and their involvement with the Roswell crash.

A man of deep compassion and humanity, Ed Mitchell has been motivated by his sympathy for his fellow townspeople: families whose lives were deeply hurt by a great conspiracy of silence; fathers who carried the weight of the death threats over their children, spoken by military authorities; threats that their wives and children would die should they ever breathe a word about what they had seen.

The day of the Roswell crash, the local press reported it. The day after the crash the official narrative cut in and the story changed. A weather balloon! But Roswell is an object lesson in which indigenous narrative survives. Before they die, many times immediately before they die, our ancestors – the *"old-timers"* – communicate to their descendants what it really was that they saw and heard.

Roswell was, in fact, one of a number of instances of group encounters with UFOs in America in the years immediately following World War II. Prior to Roswell, US Department of Defense would brief the public, and the press reported freely. The official line was, *"We are engaging with UFOs. We are pursuing them. We don't know what they are. But when we ascertain what they are we will tell you."*

Newspaper headlines of the day blazed:

Flying Saucer Found – Army Reveals Disc Picked up in New Mexico.
(Herald Express)
RAAF Captures Flying Saucer on Ranch in Roswell Region.
(Roswell Daily Record)

And then from the day after Roswell, silence. Logic would suggest that US Defense did ascertain what they were, but did not want to talk about it.

The new policy of silence was the fruit of the National Security Act, signed into law by President Truman in that same year of 1947. It laid the foundations for the establishment of the CIA, with a hierarchy of classifications that can keep even presidents well out of the loop, if necessary. Among other things, the National Security Act classified all official UFO research. It concretized the shift in policy from a promised disclosure to a pattern of silence, debunking and death threats. And the people of Roswell were among the very first to feel the weight of that shift. Ed Mitchell saw first-hand the cost of the 1947 policy and became a tireless advocate for its repeal. As Dr Mitchell speaks to the gathered press in 2001 this policy of denial has been in place for fifty-four years.

What I am suggesting is that it is now time to put away this embargo of truth about the alien presence. And I call on our government to open up – as other governments have...

For any hearers new to the topic, or sceptical regarding the UFO phenomenon, Dr Mitchell highlights the most grounded and objective aspect of this field of study.

"Read the lore..." he says. By *"the lore"* he means the report of those involved; the testimony of eyewitnesses, the current indigenous narrative. *"Only in our period do we really have evidence... Read the books, read the lore concerning crash retrievals*

and start to understand what has really been going on because there is no doubt we are being visited..."

More than once through the years, investigators asked the late Dr Mitchell whether NASA was covering up knowledge of extraterrestrial contact. Dr Mitchell's words were always carefully chosen. He would say, *"Officially NASA does not know anything about [extraterrestrials and the UFO phenomenon] because that is not their mission."*

Rather than attack NASA, Ed Mitchell's ire appeared to be reserved for US military who had threatened his fellow Americans and enforced the government's policy of denial. He believed that the classification of ET contact has kept humanity from taking its proper place among a community of space-faring civilizations, and has also deprived human society of the incredible benefits of technologies availed to us by ET contact.

Consider this: Along with every other astronaut, technician, administrator and staffer on the Mercury and Apollo programs and throughout NASA, Ed Mitchell was bound by layer on layer of official secrets laws. We can be confident therefore that there was plenty he was not allowed to say. Considering what he was allowed to say, you have to wonder what else Ed Mitchell would have longed to say, had he been given the freedom.

Amid all the speculation, report, debunking, theorising and ridicule, the one thing Dr Mitchell believed would cut through all the noise was the physical evidence of materials retrieved from UFO crashes. A lot of information was already out there in 2001 in the form of a number of books. Hence his appeal to the public to read them. But what he longed to see was the vital addition of an official government acknowledgement. Sadly, Ed Mitchell did not live to see it. He passed away on February 4th, 2016.

The very next year in December *The New York Times* released videos of an incident in 2004 in which the aircraft-carrier *USS Nimitz* deployed two pilots to engage with two Tic Tac shaped craft. These UFOs – or UAPs to use today's nomenclature – exhibited the classic properties of UFOs: unbelievable speed, physically *"impossible"* manoeuvres, and animal-like agility as they effortlessly evaded the F18 Super Hornets despatched in pursuit. The newspaper also uncovered a secret unit within the Pentagon whose remit was the investigation of UFOs.

Two years later in 2019, the US Navy released three videos and briefed the press regarding the 2004 *Nimitz* incident, and two other similar incidents in 2004 and 2015. All of a sudden, official personnel at every level were acknowledging what many had long suspected; namely that in the seven decades since the embargo of 1947 US Defense has continually been engaging with UFOs. Not only was the acknowledgement made by senior Naval personnel but we, the public, were allowed to watch the Navy's video footage of the encounter, on national television and all over YouTube.

In 2020 the Pentagon finally departed from the well-worn script, to acknowledge that for the seventy-two years since the legal classification of UFO research, a Pentagon department in fact has been in place. *The New York Times* had it right. The unit was called the *"Advanced Aerospace Threat Identification Program"*. In case the word *"advanced"* isn't clear enough to fully reveal its remit, in 2019 a spokesperson spelled it out for us. Its secret work has been to analyse physical materials from what may be crashed UFOs. The spokesperson who broke this news on national television was the man responsible for running the department from 2007 to 2017, Luis Elizondo.

In this new light, it is small surprise therefore that by the 1990s US Defense had been able to provide forty to fifty personnel for Professor John Mack to interview regarding their close encounters and abduction experiences. Given this late admission, one might reasonably ask, after seventy-plus years of engagement and investigation, what has the US Navy concluded about these UFOs? What are they? Where are they from? Apparently, the US Department of Defense doesn't know. But they will surely let us know as soon as they find out. Suddenly we're back to the pre-1947 policy. I have to ask: how much longer can that non-explanation continue to be wheeled out and received with any degree of credibility?

Perhaps that is why in 2020 some more concrete language has appeared from official sources on the public stage. Within months of these disclosures from the Pentagon and the US Navy, a consultant to the Pentagon, the astrophysicist Eric W. Davis, gave an extraordinary interview to *The New York Times*. Eric Davis has the interesting responsibility for providing regular briefings to America's Department of Defense concerning the UFO phenomenon. As recently as March 2020, Davis reported he had provided a briefing for the Office of Naval Intelligence Unidentified Aerial Phenomena Task Force concerning materials recovered from *"off-world vehicles not made on this Earth."*

That is quite a specific form of words. Between 1947 and 2019 such an assertion would have been met with an emphatic denial from official sources, or at the very least an official alternative. For instance, when Ed Mitchell spoke out in the way he did, NASA felt the need to meet the press and say, *"Dr. Mitchell is a great American. But we do not share his opinions on this issue."*

By contrast, in 2020, serious figures around the world have stepped forward to confirm the credibility of Eric Davis' statements to the press. For instance the former Director of French Intelligence (the General Directorate for External Security – the DGSE) Alain Juillet spoke to the press and said, *"I was fortunate to be [in attendance] at the origin of the Advanced Aerospace Threat Identification Program's research program on UFO's in the American Department of Defense."*

In an even more astonishing turn of events, in December 2020, Brigadier General Haim Eshed, former chief of Israel's military space security program, met the press with a statement acknowledging not just *"off-world"* technology, but asserting that covert layers of government are aware of an *"intergalactic council,"* awaiting the right time to self-disclose.

In the USA, within days of the Eric W. Davis interview appearing in *The New York Times*, Chris Mellon, Deputy Assistant Secretary of Defense for presidents Clinton and George W. Bush, appeared on CNN to make the following comments:

"I know Eric very well... I was present at [Eric Davis'] briefings on The Hill... It is an issue that should be taken seriously."

To my amazement as I sat watching, Mellon went even further. Without any prompting Mellon slapped on to the table the controversy and coverup surrounding the famed crash of 1947 at Roswell, New Mexico. He said this:

"Curiously... on Father's Day [June 21st, 2020]... President Trump indicated on air, while being filmed, that he did have classified information about Roswell, New Mexico. When asked by his son about declassifying it he said, 'I'll have to think about it.'"

"I don't know what about Roswell could be classified or interesting – other than that one particular issue."

He meant the crash and the retrieved materials.

"There is a lot of new information coming forward and I think this is a topic that the Oversight Committee should take seriously and investigate."

"In some of these cases we have information from multiple radar systems, infrared systems, multiple naval personnel, on the ground and in the air. And we're tracking these objects performing manoeuvres that clearly indicate they're under intelligent control. They're responding to our aircraft. They're out-manoeuvring them and they're doing things that are far beyond any capability we possess. This is not unique to the United States... Our military personnel have similar experiences when they are deployed abroad – in the Middle East, in Afghanistan, and elsewhere. By all accounts it is definitely a worldwide phenomenon."

It is not difficult to join the dots. Yet if I had been confronted with these announcements from government spokespeople before my research for *Escaping from Eden*, I really don't know what I would have done with them. I had no theological grid to accommodate extraterrestrial civilizations even existing – let alone turning up! Like many religious believers I saw the universe through a lens that permitted me to perceive God, the Devil, angels, demons, humans, animal, vegetable and mineral. Anything that didn't fit into one of those boxes would have to be denied or reinterpreted. But the drip, drip of credible reports, the abeyance of the policy of silence, and the bona fides of witnesses such as John Mack, Ed Mitchell, Monsignor Corrado Balducci, and Revd Dr Guy Consolmagno, has gradually eroded the containment power of familiar assumptions and beliefs. And for me, by the time Luis Elizondo, Eric Davis, Alain Juillet and Chris Mellon were coming to the stage, my world had already changed. The red pill of that plural word *elohim* in the Biblical texts had opened a door to a whole universe of possibilities.

As I have continued my explorations of Genesis and the many world mythologies it parallels, I have come to see the coherence of a worldview with room for a wider cosmic family. Given the scale of our universe, the conspicuous gaps in our knowledge concerning human origins, and the plethora of clues in fields including archaeology, cosmology, theology, astrophysics, paleobiology, DNA research and neurology, it really isn't hard to find the wiggle room to allow for the possibility of extraterrestrials – both in our cosmos and in our history. Yet even without the threat of death, whether by execution or by *"unfortunate coincidence"*, the admission of ETs into mainstream conversation is rarely made. I hear this sad fact stated and restated in my inbox with dismaying regularity. Some months I hear every week. Some weeks I hear every day from people who tell me things like this:

"I asked a question in my Bible study group about ETs and I was treated as if I was either stupid or mentally ill."

"I am a priest involved in pastoral responses to paranormal

phenomena. A number of people have reported close encounters to me. I mentioned this to my bishop and he told me never to speak of it again."

"I told my family about what happened to me twenty-five years ago when I was eleven years old. It has traumatised me and changed who I am. They couldn't stop laughing. This is why I tend not to acknowledge my experience with other people."

Humans are social animals. We are hard-wired to fit in and win approval. It is how we learn to walk and talk. Consequently, peer pressure and the fear of ridicule is more than enough to silence most of us. I say this without any judgement because this was my story too.

At the age of seventeen I had a powerful spiritual experience and made a life-changing decision to follow the teachings of Jesus Christ. It was a pleasure to discover a group of friends willing to show their vulnerability and pursue something beyond themselves. But there is a funny tension in the world of faith between enquiry and conclusions. The fact is most communities of faith are built on a certain canon of conclusions. To stray too far beyond that canon is therefore to undermine your credibility and usefulness to the group. Approval is best maintained by confining yourself to the mantra of the group's shared conclusions. I learned this lesson early on as a young Christian.

When I was eleven my parents introduced me to the blockbuster of a book *Chariots of the Gods* by Swiss writer Erich von Daniken. In the decades since, the book has enjoyed no shortage of detractors. But for me it identified with great clarity the gap in our ability to explain ourselves as a conscious, intelligent, and technological species. It also shone a spotlight on to sufficient evidence to raise the possibility that our ancient ancestors had intersected with interstellar visitors. Furthermore the book named senior researchers in the field of DNA research and astrophysics, figures such as Francis Crick and Carl Sagan, who had argued the view that life on Earth had originated elsewhere in the universe and had landed in media carrying genetic coding, sometime in the distant past, on our hospitable and watery planet. But what *Chariots of the Gods* did not do was settle the question of where we all came from. It just moved the question off-planet.

After long reflection, my wanderings led me to investigate the Jesus of the Gospels. In him I found a credible figure with valuable teachings. My reasoning about Jesus, however, didn't settle my questions about the cosmos. How did everything begin? What was or is the Cosmic Source? Is the cosmos populated? Do we share the cosmos with ET species and, if so, are we related? My appetite for answers to these questions was as keen as ever. For those reasons it was with some excitement that, at the age of seventeen, I stumbled across some cryptic sayings made by Jesus in the Gospel of John.

In the sixteenth chapter of the gospel I came across a scene in which Jesus explains to his Apostles that there are some other really important pieces of information he would love to share with them about life, the universe and everything. However, even after knowing him most of their lives (some of them) and even after three years of daily teaching, Jesus didn't think they would be able to handle the information he wanted to pass on. The Holy Spirit would have to lead them on that adventure after his time. Given that Jesus had said some pretty far out things during his three years of teaching, I had to wonder what the nature of this hard-to-process other information could possibly be.

Then in the tenth chapter Jesus says this: *"I have others who are not of this fold. I must bring them also so that we can be in unity."*

Others? Not of this fold?

Reverend Dr Guy Consolmagno argues that in this passage Jesus is referring to unnamed extraterrestrial beings – presumably the ones we should be ready to embrace *"sooner than anyone anticipates."* As a seventeen-year-old, I saw that possibility right away, and put the idea to my Christian friends at school one day. Their reaction was emphatic. *"Paul, if you're going to be a Christian, you'd better forget about your theories about the universe! OK? Get with the program!"*

So I did. So much so that I became a preacher, a church-planter, a church-doctor, a theological educator and an Archdeacon for the Anglican Church in Australia. But like many others within the program, I found that my questions remained live. Somehow, alongside all the orthodox conclusions familiar within the Christian community of faith, I always found plenty of room for questions about the universe and our many anomalous experiences as a human race.

That's why today it is with a great measure of empathy that I receive and read messages from people who have journeyed for decades with their own body of questions. Questions about their own anomalous experiences. Questions many have carried in silence. It is why I count it a privilege to be asked questions like these:

"What happened to me when I was fifteen? Have you heard anything like that before?"

"How do I live a regular life with the memories I am carrying and can't deny?"

"What do I say to my bishop who has asked me never to talk about the things my parishioners are experiencing?"

"If the elohim stories of the Bible are really about aliens and not about God at all, then where does Jesus fit into the picture? Did Jesus know about aliens? And if he knew, how did all that knowledge get forgotten? Or was Jesus an alien? Or didn't he really exist? Was Jesus invented to create a convenient religion for the Roman Empire? How does it all go together?"

In the wake of all these questions something has happened that I never expected. When first I wrote *Escaping from Eden* and began appearing on shows like George Noory's *Coast to Coast* and Sean Stone's *Buzzsaw*, and when I first started generating video content on *The 5th Kind TV*, I somewhat expected that popular demand for me as a pastor would fall off rather dramatically. Instead, personal requests for my services as a pastor and personal coach have exploded out of my little corner of Australia and opened up all around the world. Today I am sharing the journey with a great many as they navigate the confusion that so often results as an old worldview trips over itself and another one takes shape. It's why on this particular morning I am sitting in my study in Canberra, Australia, with my laptop ready, waiting for a call from Reuben. He is in Lithuania, sitting in his study, wondering how aliens in the Bible are going to impact his work as a Baptist missionary.

Chapter Seven

Lost and Found

Kaunas, Lithuania – 2020

"Paul, I totally follow your logic. You've convinced me. The Elohim are the Powerful Ones. The Powerful Ones are the Sumerian's Sky People. And the Sky People are extraterrestrials. I get that. But I am struggling here. I have been on the field a long time and my message has been that the Bible is God's book, telling the truth about the world. I have always thought of Jesus and the New Testament as taking up where the Old Testament ends."

"This year has been a funny year and it's given me a bit more time to go over some of my questions. Your channel has really given words to a lot of my own thoughts over the last couple of years. A lot of my questions get resolved by the interpretation you're offering. The problem is it upends a lot of my old thinking – the Bible is God's book and Jesus is the punchline. If that's not true then I don't know what the Bible is anymore. It leaves me wondering if I have been wasting my time!"

Reuben is a missionary and a pastor. For more than twenty years his parish has been in Lithuania. It's a predominantly Catholic country but Reuben operates there happily as a missionary for the Baptists. Culturally, Lithuania is a country of many layers and some of the country's folklore has pre-Christian, Norse undertones. These undertones echo with the memory of ancestors bumping up against non-human neighbours from off-planet. Intrigued by these mythologies, Reuben tapped in a Google search and found the Paul Wallis Channel on YouTube. And that was his red pill. Now, a few months down the track, he is wrestling the same wrestle I have wrestled as his worldview shifts and reforms.

I liken this wrestle to the stages of grief, identified by Swiss-

American psychiatrist, Elisabeth Kübler-Ross. When a discovery upends our familiar world and makes it impossible to return, we go through the same processes as when we lose, or prepare to lose, a loved one.

First is denial. Sometimes the first reaction I get to my message is anger. *"No that's not right! Elohim is the three persons of the Holy Trinity. Aliens are a demonic deceit or a government conspiracy! You're blaspheming and leading people away from the truth."* Etc.

Second is Bargaining. This is where we try and massage and contort the new information to see if we can force it to comply with our old worldview.

Third is Depression. Reuben's work has been wrapped around a worldview that he can no longer support. His sense of significance has been supported by a very particular view of God that he now sees is peppered with illusions. Can he really let go of all that and understand himself and his life differently? To think he has been mistaken about so many things, or to think that his theological educators have been so lazy as to cordon off whole sections of what our forebears were able to discuss, and then to think that we all have been left on the outside of vital information known to higher levels of government and intelligence... well all that leaves a person feeling not very special. There is a sense of deflation and *"What do I do now?"*

Then comes Acceptance. You find the good. You come to terms with the journey. Change is inevitable. Growth is positive. You are making your way through life and becoming more conscious and more understanding of your own story and the story of all of us.

In his own words, Reuben is telling me that he has moved from bargaining to depression and is now on his transition to acceptance. As we talk together on Skype, my goal today is to help him make that crossing. So, to consolidate that step, we begin with Reuben's workbook, the Bible, and take a look at how it came to be assembled.

"OK, Reuben, let's look at some history. First off, whenever we hear Jesus and the Apostles quote 'the Scriptures' they are quoting a Greek book – the Septuagint. Sometime between the C3rd-C2nd BCE – a team of translators brought the Hebrew Scriptures into the international language of the day. While they were doing that, they added some newer writings which seem to bridge Hebrew culture and Greek thought. In that collection of writings, you have new ideas about human consciousness, its pre-existence and its survival of death; ideas of other entities, angels, demons, hyper-intelligent and extra-dimensional beings; notions of inter-dimensional contact. So, all of that was in the air when Jesus came and began making his input. That's the thought-world in which Jesus and the Gospels emerge. Right?"

So far, so good.

"Now it's worth noting that when Jesus and his apostolic writers quote this Greek book, they almost never take the plain meaning of it. They always come away with metaphorical or esoteric meanings. Don't miss that. Then in Acts 15 in the New Testament we have a record of Jesus' brother presiding over a council that declares that the Church is no longer bound by the Hebrew books. So don't miss that either!"

Lightning fast, we then run through a thumbnail sketch of how the Hebrew Canon came together.

To go right back to the roots, *Genesis 1-11* give us the stories of beginnings – the summaries of the Sumerian stories of Sky People – the stories that Abraham and Sarah grew up with in Ur of the Chaldees before leaving to found their own tradition.

Sometime later these Sumerian-based stories get included into the books of Moses – the Pentateuch – The Torah – the first five books of the Bible, Genesis to Deuteronomy. The world described in these books from the C7th BCE is a henotheistic world. That means there are many *Powerful Ones* or *"gods"*, but the devout Jewish reader should follow only this one –

named *"YHWH"* – hence the Ten Commandments and Joshua's command to forget the *"other Powerful Ones"*.

Within the next century or so the Pentateuch becomes the introduction to a collection of writings called the Deuteronomistic History, which is a sequence of historical stories interwoven with moral interpretations. These books, *Deuteronomy* to *II Kings,* have been edited together into a single volume intended to unpack the laws by which the ancient Israelites were governed. It's still a henotheistic world and the violent competition among the Powerful Ones is what drives many of the Deuteronomistic dramas.

In turn the Deuteronomistic History then becomes the introduction to a body of literature which included some new history with interpretation, some poetry, songs and proverbs, and writings from prophetic sources. That great diversity of literature becomes the Hebrew Canon.

Here is where the plot thickens. Because by the end of the C6th BCE someone has undertaken a major edit to harmonise the diversity of writings into a single work with a unified theology. The multiple *elohim* of the Pentateuch have been airbrushed out. Almost. The competition among the *elohim* of the Deuteronomistic history has been reframed as stories of conflicts on a human canvas. It becomes worship *vs* idolatry or obedience *vs* sin. Monotheism is the word on the street and Judaism's every cultural debt to Sumerian culture has been written out of the story, and every parallel disguised.

Many scholars believe this edit was done during Israel's painful subjection to Babylonia – a daughter culture of ancient Sumeria. Perhaps as a consequence of that unhappy context, there is in the retelling of the old, old stories a palpable undercurrent of rage. *"Our God is real. Yours is a fake. We will be eternally vindicated. You will be eternally shamed."* This Us and Them narrative undergirds the whole editorial framework for the Bible as we have it. And it shows.

"So, Reuben, when the early Church made the decision to glue the Hebrew Canon on to the writings of the Apostles of Jesus, they brought that same basic framing into Christianity. With that as the introduction to Jesus, what Jesus was on about gets framed as a religion of us and them; worship vs idolatry, obedience vs sin, church vs world, heaven vs hell."

"All that fear and hate, crime and punishment – in fact, the whole framework of authority and obedience – makes for a form of religion very amenable to political powers hoping for a passive and compliant people. Consequently, it wasn't too long before the Roman Empire found a way to assimilate Christianity. Powerful people began putting up grand buildings to house the Church, and the Emperor awarded the Bishops imperial rank and put senatorial purple on their robes."

"Of course, the transformation would hardly be complete without a makeover for Jesus as well. If you want a snapshot of that, if you go to Ravenna in Italy you will find a beautiful portrait of Jesus in the Archbishop's Chapel. The mosaic shows Jesus in the finery of Roman military uniform. This Jesus is, evidently, a soldier of the Empire, humbly taking imperial orders, and armed so he can mete out imperial justice. In this picture the very ideas of a 'good Christian' and a 'good Roman citizen' have become indistinguishable. Hey presto! Imperial Christianity!"

I find that Reuben is in fact very familiar with this whole tract of religious history. With roots in a dissenting tradition within the Christian faith, he has a good nose for detecting the creeping vines of institutionalism. His faith has always resisted feudalised religion and favours the more democratic vision of a priesthood of all believers. So thus far, he and I are on solid ground together.

"I think you would agree with me that the Gospels look like they have crystallized a good time before that imperial takeover. And I reckon that what's in them is so at odds with that passive religion of worship, obedience and good citizenship that it's ridiculous. I don't think the Jesus of the Gospels teaches quiet, trusting compliance.

'Have no leaders,' he says. 'You are all brothers.' He's a bit of a rebel. Now, Reuben, I've listened to your preaching online, and that's the Jesus I hear about from you. If you're helping people to follow the teachings of Jesus in the Gospels, I think you're on a strong wicket."

These are all conclusions Reuben has already reached. His work as a missionary preacher over more than twenty years has required him to spend half a lifetime in the Gospels and for him the clear blue water between the Jesus of the Gospels and the tradition that emerged in his name is not hard to see. The distance that Jesus placed between himself and the Hebrew Canon has also been part of Reuben's theology since college. What is a bit fuzzier for him is the question where Jesus stood in the world of ideas back in the time of his preaching. This is the stepping stone I will need to direct Reuben to today.

Graeco-Roman culture was a veritable melting pot of international thought. The philosophies of India, Asia, Africa and Europe all flowed into the mix. Better than anyone, the Greek philosopher Plato managed to distil a coherent set of ideas that made sense of a world's worth of philosophical and ancestral knowledge. It was a fusion which he expressed in his books through expositions of Socrates' thinking – though where Socrates ends and Plato begins is a bit opaque. It was with this synergy of international thought as a backdrop that Greek Christianity first emerged.

"Reuben, to see where he fits in the thought-world of that time you need to take a look at how Jesus relates to Plato. For instance, when Jesus describes his own journey of consciousness, it absolutely parallels what Plato says about all of us.

John's Gospel says, 'In the beginning was the word and the word was with God and the word was God. He was with God in the beginning.'

Plato says we all begin that way. Before this material life, our consciousness is part of the source consciousness, the Divine consciousness."

"John's Gospel then goes on, 'The Word became flesh and dwelt among us. We beheld his glory... full of grace and truth.'"

"Plato says we all come into this material world to wrestle with that exact same fundamental question: 'Can we, in this material life, do consciousness and intelligence, love and harmony (what John calls grace and truth) as a society of individual consciousnesses, each exercising free choice?'"

"Then at the end of John's Gospel Jesus says, 'Father I am about to return to you – back to the glory we enjoyed before the foundation of the world.'"

"Plato taught that this progression is actually the fate of all of us. That's why Plato's followers – including his Christian followers – did not fear threats of hellfire. At the end of this life they expected some kind of a life review and then for their consciousness to move on to the next thing."

"For me, in the light of Plato, I look to Jesus and see him more than I ever did as a model. He shows us what is possible. And what I am saying is what many of the early Church Fathers believed – people like Justin Martyr, Clement of Alexandria, Origen and Marcion. It's why they hinted at the idea of an Old Testament of Plato and a New Testament of Jesus and his Apostles. It's why they quoted Apostle Paul and Plato side by side."

Reuben has some recollection of Plato from his theological training of thirty years ago. Yet just like myself, and like so many whose academic studies have required them to engage with the intellectual vastness and importance of Plato, Reuben came away from college dimly remembering a few of Plato's more generic contributions, but having completely forgotten his most world-changing ideas. When the Church Fathers I mentioned affirmed Plato and quoted him as an equal, or more pointedly, as a superior intellect to the Apostle Paul, they were affirming a philosopher who championed some powerful and eye-opening beliefs which were soon to become taboo in imperial Christianity:

- The *"children of God"* who modified our ancestors to increase our capacity for intelligence and consciousness. (This idea resounds throughout the mythologies of the ancient world.)
- The holographic universe and the non-divine *"Craftsman"* (*demiourgos*) who translated the codes or *"forms"* and downloaded them into the physical manifestations we are all familiar with. It's like *The Matrix* in ancient Greek!
- Our cosmic neighbours, longer lived, more intelligent, and with an advanced knowledge of outer space and residing on their *"islands in the sky"* – what the Drona Parva of the Hindu Vedas described as *"cities in the sky"*. Today we might call them space stations or *"Unidentified Aerial Phenomena"*.
- The succession of civilizations on Earth. Every few thousand years, he said, the movement of objects in space would trigger extinction level events effectively resetting civilization to a virtual zero.
- The conscious universe – or at least the idea that consciousness preceded and generated this material realm. This idea echoes in C21st quantum research.

"Reuben, did you know that Plato described Earth as a globe, floating in space? He even had a fair go at describing what planet Earth looks like from a distance. He said it was a surprisingly swirly pattern of blue, white, green and gold. To my mind that description is pretty impressive, given he made it two and a half thousand years ago. I mean, if you look at 1960s popular science fiction, planet Earth is always shown as a spherical version of a textbook map, with land of gold and sea of blue, and not a cloud in sight. That shows you just how far-sighted Plato's vision of Earth really was."

Think about that. How could Plato have conceived of that view of the planet? Some might say I have cherry-picked Plato's best bits. There are other moments in his books where he appears to repeat another convention – the classic vision of a hollow, flat Earth beneath a domed canopy of sky. However, Plato gives voice to at least three sources of information. One source has provided him with the view we recognize today, a planet in space, a swirl of those same colours. How?

Our culture didn't see our planet that way until the 1960s, from photographs taken on the Apollo missions. It was in 1966 that the cameras aboard Apollo 8 took a now iconic photograph of planet Earth, partly shadowed by the moon. The picture was called *Earthrise* and it was an absolute revelation. In an instant how we all conceive of our planetary home was transformed. You can see this incredible epiphany reflected in the 1960s TV series, *Star Trek*. When the show opened in 1966, planet Earth and similar planetary counterparts were portrayed as classic map-style globes of gold land and blue sea. However, by the third season in 1969 Earth had changed and transformed itself into the familiar swirl we now know and recognize.

In 1967 the NASA communications and weather satellite ATS-3 took the very first photograph of the whole planet in its near spherical glory, suspended in space. It shows a beautiful swirl of white, blue, green and gold. It seems to me this was an obvious moment to reopen the pages of Plato with a good deal more wonder and curiosity as to where, two and a half millennia ago, this ancient scholar got his information from.

In fact, Plato tells us. These were his three sources:

Plato's first source is his cool, systematic method of patiently applying logic to things we can all observe. Plato called this philosophy. We might even call it science.

Secondly, Plato identifies knowledge derived from ancient Egyptian priestly sources. This knowledge was passed on to an ancient Greek legislator – a real historical figure called Solon. The information then passed down the generations of Solon's family, finally reaching Socrates via one of his pupils. Plato then curated that information for posterity.

Thirdly, Plato names information about extra-dimensional and spiritual phenomena, derived from what we would call a psychedelic experience, induced by a fermented tea called Kykeon. The Kykeon ritual was the centre point of ceremonies undertaken in the privacy of the Eleusinian Complex of Athens.

Several days of quiet preparation and fasting were required before the Kykeon could be ingested – and this was done with strict supervision. Although it's hard to be certain exactly what Kykeon was, there is some consensus among today's researchers that it was most probably a cocktail of fermented barley and mint, infused with pulverised ergot – a psychedelic fungal extract. Both the process and the experiences it generated sound remarkably similar to the psycho-effective results of psilocybin or ayahuasca ceremonies. When people today describe the effects of psychedelic teas, they speak of transcendental experiences, or encounters with extra-dimensional entities. They talk about spirit guides who show them other dimensions or explain the guidance that will prepare them for the phase of existence that follows this one. Experiencers often speak of being forced to confront their fears of death and overcome the diminishing power of fear itself in their life on Earth. In short, they are made to wrestle with the heaviest and deepest issues of their own soul's journey, often returning from their wrestling with a determination to take on the challenges of life with renewed confidence and enthusiasm. That is exactly how Plato described the experience of the Eleusinian mysteries. Plato expresses all this in the voice of Socrates, yet he describes it all so vividly and with such internal detail that we get the distinct impression he knows about these things first-hand.

Plato's writing is vital for so many reasons. It lies at the foundations of so much of Western thought. It also shows us what vision of God and the universe would have framed the Christian story of Jesus if thought-leaders like Justin Martyr, Clement of Alexandria, Origen and Marcion had succeeded in creating an Old Testament of Plato and a New Testament of Jesus.

"So, Reuben, if you really want to unplug from the matrix, I reckon Plato is going to be the key for you. He is the White Rabbit!"

Perched in front of his iPhone in Lithuania, Reuben is massaging his scalp to help him process this overwhelm of information.

"Wow, Paul, I am really ready for that White Rabbit! I think I knew most of your Biblical material," he says. *"But Plato... I can tell you we did read Plato in Seminary, but I guess mainly to quote him for essays on Christian orthodoxy. To be honest, I think I only ever read Plato quotes on handouts. I don't think I ever sat down with a whole Plato book. We just didn't have time. I certainly don't have any memory of reading what you just said. Where are you getting all this?"*

"Reuben, next time you take a retreat, sit down with Plato's Phaedo *and* Timaeus *and* Critias. *Get them through my website. Take them to the beach or up into the mountains with you. I can tell you, what's in those books will absolutely blow your mind!"*

Chapter Eight

Times and Places

Balmain, Sydney, Australia – January 2020

The Greyhound bus has brought me to some familiar streets. I love this part of the journey. I always remember the sense of adventure the first time I landed in Australia more than twenty years ago. I could sense that new discoveries awaited me here. And they did. Here I met my, now wife, Ruth, and together we began a whole new journey of exploration. Today I am in Sydney to make some fresh discoveries. I am meeting a small gathering of researchers and experiencers of ET phenomena.

It's taken three and half hours to get here from Canberra, but I have made the trip in the company of *Phaedo* and *Timaeus and Critias* – the two books of Plato I recommended to Reuben. To be fair, Plato's writing style has taken me some time to get used to, but I have found it well worth the effort. There is so much in those books to expand my thinking, and so much to intrigue and mystify.

For instance, when Plato speaks of extraterrestrial *"Children of God"* who or what does he have in mind? He teases us with that language. In fact, his choice of words echoes the language of Genesis 6 – the *"benei elohim"* who came and hybridized with human females in the story that precedes the Great Flood. By contrast, Plato's associations with the *"Children of God"* are positive. For him they were benevolent beings who nurtured our development, heightened our intellect and helped develop our state of consciousness to better enjoy the potential of life on Earth.

Another mystery. Who were the mysterious *"others"* living on islands in the sky with so much advanced knowledge of the wider cosmos? The Mesopotamian texts call them *Anunnaki* or

Sky People. Genesis called them *Elohim* – the *Powerful Ones*. The *Popol Vuh* describes them as *Those who Engineer*. The Vedas, the Nordic and Greek texts tell stories of *kings* and *gods*. Babylonian legend refers to the *Apkallu*. Ancient narratives from the Efik people of Nigeria name them as *Abassi* and *Atai*. The Dogon People of Mali, West Africa say their prehistoric tutors came from a planet orbiting Sirius C. The Egyptian and Mayan books of the dead both point to stars in the constellation Orion. Aboriginal Australians and many Native American tribes point to the Pleiades and say that those who nurtured their cultures came from there. Somehow, Plato achieved a bird's-eye view of this great canon of ancient knowledge from around the world – all of which raises an obvious question: How can we possibly have forgotten what he and so many cultures have remembered?

As I begin to make my own tour of the world's ancestral narratives, as Plato must have done two and a half millennia ago, I notice that the same specific regions of outer space keep cropping up. When identifying the original home of our prehistoric visitors, ancient cultures repeatedly point to three regions among the stars: Sirius, Orion and the Pleiades. Intriguingly, the Bible's oldest book, Job, connects all three.

The book of Job is a bit of a mystery. Some scholars believe it to be an Arabic book, which somehow migrated its way into the Hebrew scrolls, having been made over into its current form. It is one of a number of texts that clue us to the presence of a Sky Council – a cabal of advanced species, governing over project Earth in our distant past. In the 38th chapter the book of Job says this:

Can you bind the chains of Pleiades? Can you shake off the cords of Orion? Can you take charge of a family of stars in their season? Can you lead Sirius with her satellites?

The chains of Pleiades? The cords of Orion? The leadership of

Sirius? The language is all about power exerted over humanity from these three regions of space. The writer asks the character Job if he is powerful enough to turn the tables and take charge over them rather than vice versa. Is this a question about overcoming the influence of the stars of planet Earth – *i.e.* overturning the seasons? Or is the question about chains, cords and who's in charge of who, actually a question of overcoming the power of the visitors from these particular stars? It's an odd question and an odd coincidence that those are the three constellations named.

At the gathering of researchers and experiencers in Sydney, I find myself in an animated and eclectic circle of people. It's my privilege to be among them. Over lunch I fall into conversation with a Cherokee friend who is happy to fill me in on his own story. Chad's patient storytelling introduces me to the prominent place of people from the Pleiades in the ancestral story of many Native American tribes. For instance, the Navajo of New Mexico, Arizona and Utah identify the stars of the Pleiades as being suns similar to our own. Indeed, in terms of distance the Pleiades are the closest stars to our own sun and may have been formed in the same process. This, in itself, hints at the possibility of similar life in that part of galaxy. The Lakota people of the upper Mississippi region claim that their ancestors originated in the Pleiades and that when humans die, their consciousness returns to the Pleiades.

The Lakota belief about our soul's afterlife finds an intriguing echo on the other side of the world in East China. At the Xiaoling Mausoleum near Nanjing, lie the remains of Emperor Zhu Yuanzhang – founder of the Ming Dynasty. At his death in 1368 CE the emperor's servants interred his body in a complex of tombs. When viewed from the air the arrangement of tombs aligns perfectly with the visible stars of the Pleiades. Evidently, Zhu Yuanzhang anticipated a journey after death, just like the one anticipated by the Lakota people of North America.

Chad first learned of this Pleiadean connection from his father and Chad passed it on to me.

"Our ancestors saw egg-shaped craft arrive from the night sky. The people who emerged from the craft were tall and powerful beings. For a time, they lived among our ancestors and showed us how to farm; taught us about health and sanitation; how to cultivate some plants for food and others for medicines. They taught us about the seasons and to live in harmony with the land. After a time, they returned to their craft and made their journey home – a place among the Pleiades."

"In my family though we know that our ancient helpers are still among us. From time to time their craft still visit us. My own family carries stories of healing imparted by these people. That is why I have never feared visitors from the stars. I always assumed they were friendly towards humanity."

Chad is not the only Native American at this gathering. As we walk around a park flanking the waters of Sydney Harbour, he introduces me to Dean, another friend with Cherokee roots. His experience has left a different kind of mark. Dean works in the world of environmental management. In 1999 he was managing a site in regional New South Wales. It was just after lunch and the team members were arriving back from their breaks and were taking up their places around the site. At precisely 2pm the sky darkened. Dean and his colleagues looked up expecting to see a dark storm cloud. Instead, hovering directly above their site was an enormous, saucer-shaped craft. The next thing any of the team remembered, they were waking up, each lying or crumpled where they had fallen. It was 4pm and not one of them had any recollection of the previous two hours.

"Dean, do you have any physical evidence of what might have happened during that missing time?"

"I have a photograph," he says. *"When we all came to, we could see that one of our team was quite distressed. We gathered round her and saw that she had three marks on her upper arm – three raised lines in the shape of three fingers. They were raw and red, like a burn mark,*

*or like an allergic reaction to something that had touched her bare
skin. So, I took a photograph of her arm and I've kept it."*

"How long were these finger marks?"

"About eight centimetres."

"About eight centimetres?"

I tried to imagine the disorientation of the scene, the oddness
of those scars, and the anxiety that must arise from not knowing
what had just happened.

*"Dean, have you ever thought about doing some regression therapy
to see if any other memories surface?"*

Dean is emphatic.

"No. I don't think I want to remember."

I am deeply grateful to Dean for entrusting me with his story.
In telling you I have changed some inconsequential details out
of respect for his privacy. It is a very vulnerable thing to share
an experience that you are still in the process of understanding
and I can easily understand Dean's reluctance to retrieve more
memory of the encounter. After a trauma our brains often block
out the recollection of experiences it might be too painful to
remember.

I now wonder if that fact might apply to whole cultures too.
As I have read the repeating stories of our world mythologies and
found story after story suggesting past, traumatic intersections
between humanity and visiting species, the stories read like the
flashbacks of a patient with amnesia, gradually retrieving the
lost moments of a disjointed story. That three fingered scar.
Something about it bothered me. I would want to know what it
was. If I had lost time, I would want to remember. Wouldn't I?

As I sit in the air-conditioned safety of the Greyhound bus,
I contemplate the dusk-time bush as it blurs my way from
Sydney back to the Australian Capital Territory. Staring through
the window into the darkening landscape I can feel something
simmering at the back of my mind – the stirring of some vague
memory.

It's 1985 again. I am twenty years old and am standing in Great Pulteney Street in the city of Bath. A very attractive young lady has just seen me and called out my name. We have seen each other just recently. Where was it now? My recollection is strangely foggy. As we continue talking, she references places and events from the last couple of days. But my memory is a blank. I have no recollection of the places, events or days – or her. Desperately, I try to maintain the pretence that I know exactly what we're talking about – but ultimately without success. We both leave the conversation, frustrated and horribly embarrassed.

As the Greyhound passes the turn-off for Wollongong another half memory jostles its way to the surface. I am in my flat on a bright sunny day in 1985, studying the Gospels as I prepare for my first job in the world of ministry later that year. The entry-phone buzzes.

"Hi Paul. It's Julie. Just passing by. Thought I'd call in for a coffee."

Julie and I already know each other and she has remembered a lot about me from our previous conversations – which is nice. We chat idly for a while over a pot of freshly brewed coffee – a new brand I am trying out. Julie mentions that she is living in a house in the village, just round the corner from the general store. I know the house. It is notable for being one of the newer properties in the village – late nineteenth century. As she leaves, I promise to call in on her for a coffee in return. It was a pleasant catch-up. Unfortunately, it's another total blank. I have absolutely no recollection of that girl. None! This time, however, I can follow up. I know Julie's address. I know the house. Except that when I call round a couple of days later, I can see that the house is empty. Nobody lives there. Had I remembered wrong? Could I be forgetting more than I thought?

Among family and friends my memory for conversations is well-known. In fact my verbal recall is often sharp enough as to recapture conversations verbatim from years ago. I don't say that to congratulate myself. It is not always a virtue! My recollection of places and dates is also remarkably sharp. Sit me down with friends or family and a photo album, and I will be the one who can pinpoint the place, year and month for each shot. So these two blanks were highly uncharacteristic and bothered me a great deal.

I was not one to go clubbing or get myself into situations where a drink could be spiked, or a pill popped. I was a clean-living guy. Still am! So, these blanks were a real anomaly. Nothing like them has ever happened to me – either before or since. Over the years, from time to time, my mind has returned to the puzzle of these two encounters, only to emerge none the wiser for my efforts to explain these two holes in my memory of 1985.

Tonight, on the bus nothing further offers itself up. Yet something Dean said, a phrase he used, repeats itself in my mind and seems to offer a new framework for these absences. Both were stories of missing time. Could it be that my brain has been protecting me? Has my subconscious mind chosen not to recall days when something traumatic or humiliating may have happened? Is there something my subconscious might not want me to remember?

Canberra, Australian Capital Territory
July 2020

With a notepad and a glass of water I am at my desk just in time for a call from Massachusetts. Patricia is a senior professional with a background in high level scientific research for a major corporate interest. Today she is sharing with me questions she has carried regarding her own sequence of other-worldly encounters, stretching from the present all the way back to the 1970s when she was a college student in the town of Portales, New Mexico. It was in that chapter of her life that Patricia was entertained one afternoon by a group of people who, though she only met them briefly, have remained indelibly in her mind ever since. The people were tall and light skinned, blonde-haired and with a distinctly Scandinavian look. They were, she said, the most perfectly beautiful people she had ever seen.

The girl, a fellow student, was talkative and friendly and seemed to have taken quite an interest in Patricia. The two men were huge and so identical in appearance, dress and haircuts that Patricia assumed they must be twins. The twins were taciturn and appeared to have an almost telepathic communication between the two of them. At some point as their social gathering wore on, something about the situation began to spook Patricia, and she asked to be driven home. So, it was with a sigh of relief that she climbed into her own bed that night. Except, now something else was troubling her. As she lay down to try and sleep Patricia realised she had absolutely no recollection of her journey home. Nor could she explain why it was now 2am. It was all a fog.

Over the decades that followed, Patricia did her best to forget about the whole incident. It seemed like nothing. There was no substance to it. So what if the two guys were model handsome, huge, identically dressed and groomed – and telepathic? Perhaps it didn't mean anything. What did it matter if she couldn't remember the journey home? Maybe she drank a spiked drink. Perhaps she had simply slept all the way. It all seemed like nothing. Only, she could not forget it.

The name of the town where it all happened is Portales, the Spanish word for *"portals"*. It sits on Route 70 between the Cannon Air Force base in Clovis, and the Air Force base at Roswell in the state of New Mexico. Patricia always wondered if that location and the historic reason underlying the name of Portales might be connected in some way with the strangeness of all that she remembered of that episode. Now in 2020 Patricia has been able to compare notes with other experiencers and has found a different lens through which to view her decades old story.

"Paul, have you ever had a Pleiadean described to you? Do my strange friends sound anything like that to you?"

I offer Patricia an answer from my knowledge of world mythology. Certainly, there is no shortage of ancestral stories, folklore and religious texts about benign human-like neighbours who are tall, beautiful, athletic, blonde-haired, and super-powerful. Perhaps at certain times and in certain places people might have called them *"gods"* or *"angels"*. In the lore of contemporary experiencers, they are called *"Tall Whites"* or *"Nordics"*. And yes, they are associated with Pleiades.

As I lay this information on the table for Patricia to consider I find my speech slowing as I labour to keep myself from being distracted by the most vivid and persistent memory. It is yet another odd encounter – an experience I have done my best to forget. Not because it was traumatic. The problem was I could never for the life of me explain why the experience mattered – what it was about and why it had bothered me so much. Consequently, it is a story I have seldom told.

At the time, the encounter threw a clunky old spanner into the works of my well-ordered theology. What I mean is that as a young Christian believer, my orthodox faith had all the right places for God, the Devil, angels, demons, animal, vegetable, and mineral – and nothing else. I just could not work out how this anomaly of an experience fitted into any of these categories. That's why over the years I have almost succeeded in persuading myself that it really was nothing at all and is far better forgotten. Today, as I listen to Patricia's tale of unusual hosts, I am not so sure.

Chapter Nine

Anomalous Art and Secret Messages

Planet Earth – 2020

In a year of social distancing, conspiracies and lockdowns our new neighbours have offered us a wonderful and warm moment of respite. Their little boy's birthday has occasioned a beautiful gathering of friends and family to mark the happy day. It is an international event, all regulation-safe, which has brought together the cultures of many countries, Cameroon, Nigeria, Ghana and Kenya in particular. The party has naturally segregated itself for the moment into three demographics, the women at one end of the room, the kids in the middle and the men in a loud cluster around the barbie. Gathered around the firepit, our conversation among the men has fallen to storytelling.

Each culture has its own story of human origins – both the official religious narratives and the older traditional narratives carried orally through the generations. I have already told you one of the Ghanaian mythologies, and two of the Kenyan narratives. Tonight, I am being treated to a telling of an ancestral story curated by the Efik people of Nigeria and south-western Cameroon. According to the Efik people, humanity's origins relate to the arrival of advanced beings on planet Earth in the long distant past. The story begins when a male and a female entity, named Abassi and Atai, appear in a craft which hovers high above the world, like an island in the starry sky.

When Abassi and Atai create the first humans there is no sex. There is no marriage. Every individual human is engineered or cloned on the island in the sky. The level of intelligence of the resultant humans is like that of children, and they live in total dependence on their celestial creators. Their food and medical needs are all provided for directly by the advanced beings. They live on the surface of the Earth during the day but are brought back to the island every evening. In due course, however, the humans develop into a more mature species, capable of looking after themselves, which they proceed to do. Abassi and Atai realise that the time has come to release the humans from their enclosed safety to fend for themselves on the Earth. Before long the humans have learned to farm and they begin to establish themselves as a self-sustaining culture. Once they are satisfied that the humans can live self-sufficiently, Abassi and Atai decide to take some well-earned leave from project Earth. In the absence of their superiors the humans reproduce and as the population grows they migrate further out into the world.

After their extended leave of absence, Abassi and Atai return to the Earth, where they find that humanity has developed into a capable and sovereign species, ruling over their environment and developing a powerful civilization all their own. Far from feeling any sense of pride at the success of their project, Abassi is positively alarmed. He immediately perceives that his and his partner's controlling position over humanity is imperilled by the humans' rising level of intelligence, and their power in numbers. After an emergency conference, Atai volunteers a solution. Her means of keeping the rulers in power and the humans in subjection is brutal and simple. Devices will be released to spread disease among the humans. Explaining the strategy to her partner, she tells him, *"If we distress the humans mentally, make them confused and scared; and if we can make them sick physically, then they will be no threat."*

Atai's strategy echoes the pattern of *"upgrade, upgrade, downgrade"* which repeats in the Genesis, Sumerian and Mayan explanations of human evolution. Atai's logic also prefigures the social theories of Thomas Malthus in the C18th and Charles Darwin of the C19th, which identified conflict and disease as nature's response to a planet afflicted with too many humans. In the 1980s Prince Philip, the Duke of Edinburgh, put fresh words to the theory when Fred Hauptfuhrer interviewed him for *People* magazine. He said, *"Human population growth is probably the single most serious long-term threat to survival. We're in for a major disaster if it isn't curbed – not just for the natural world, but for the human world. The more people there are, the more resources they'll consume, the more pollution they'll create, the more fighting they'll do... If it isn't controlled voluntarily, it will be controlled involuntarily by an increase in disease, starvation, and war."*

It would appear then that the push and pull of debate over sustainable population levels is as old as humanity itself. It dates to a time when there would have been a tiny number of humans rattling around on the planet, the time of Abassi and Atai in the Efik narrative, the era of the *Powerful Ones* of Genesis, and the *Sky People* of the Mesopotamian stories. By all those accounts and more besides, the Sky Council which our ET colonisers comprised divided repeatedly over their conflicting agendas for project humanity. The divisive issues included:

- *How many humans should there be on the planet?*
- *How much understanding should we allow the humans?*
- *How healthy do we want the people to be? How long lived?*
- *What access should the humans have to food and medical cures?*
- *What access should we allow the humans to technology?*
- *What access should we allow the humans to us?*

All these questions rumble through the ancient narratives with an oddly familiar tone. Our battles in the C21st over GM food patents *vs* access to food and food security, corporate *vs* public access to safe drinking water, medical patents *vs* access to medication, issues around censorship *vs* medical and journalistic freedom, access to education, unaccountable corporations *vs* transparent government; all appear to be contemporary iterations of the old, old struggles dramatized in our world mythologies. These strange echoes raise the question of whether our world mythologies might be as much about understanding the present as they are about remembering the past. Do these issues recur through the ages simply because these are the kinds of areas that will always be battled over? Or is it that since the first telling of our ancestral narratives, nothing has really changed?

By the time we migrate from our animated men's huddle around the barbie into the main room for singing and birthday cake-cutting, our conversation has ranged over a great swathe of subjects, from the story of Abassi and Atai, to Bill Gates and the World Health Organization, from farmers' markets to Monsanto, from the theft of African gold to Prince William berating Africans for having too many children – and doing so in the same year he announced the advent of his third child. Given that our group has gathered from four African nations to celebrate the birth of an African baby, you can imagine that the prince's comments have not gone down well. Though we have not actually solved any of the world's problems by the end of our fire-pit conversation, we have at least enjoyed a fun discussion and it has certainly been interesting to see how widespread and longstanding are these questions concerning the health and empowerment of human beings.

In the beginning, our world mythologies put forward a more fundamental explanation as to why these same issues harried and harassed our ancestors. Simply put it was that our ancestors were governed over by rulers who were not human. It is the fundamental reason that the Sky Council lacked any kind of fellow feeling with the human beings they presided over, and why it was capable of such cynical decisions with regard to the welfare of ordinary human beings. It is why the Powerful Ones of the Hebrew stories could send their humans out to war against each other so frequently, with little or no regard for the human cost of their squabbles.

The template in which earthly government or royalty begins

with ET overlords, only to be handed over to human regents at a later stage, repeats in culture after culture, all around the world. It appears in the Biblical narrative, Greek, Indian, and Norse mythologies. The Edo people of Southern Nigeria and the Yoruba people of Western Nigeria recount the same story. They describe humanity's original rulers as superior beings who came from the skies. The Yoruba name for the first great superior is *Olodumare*. The Edo call him *Osanobua* – a name which means, *"The Almighty One who lives Above the Waters"*. Beginning with his son *Ogiso Igodo*, the descendants of Osanobua maintain their hegemony over the human race. Their title *"Ogiso"* means *"Ruler from the Sky"*. At a later stage the *Rulers from the Sky* hand the reins of power over to human successors who, not being from the sky, carry a different title, *"Oba"*, which simply means *"ruler"*.

In many narratives the handover of power from ET overlords to human kings occurs peaceably and, happily, this is the case in the Edo and Yoruba telling. In the Biblical account, by contrast, the transfer of power from YHWH to the people of Israel's first human king is an ugly mess.

If any of this is true, if our ancestral narratives are carrying ancient memory, then the implications are far-reaching. Consider for a moment: What would it do to the self-perception of a human race governed over by entities demonstrably superior to themselves? What would it do to humanity's fundamental concepts of leadership for their first experience of it to be modelled by beings with absolutely no fellow feeling for them? It is easy to see how the experience of such cold governance, devoid of any human empathy, would leave a mark in our collective psyche as a species, and establish an unholy template for royal and government power for the ages to come. Of course, the $64,000 question is, *"How much has changed in the aeons since?"*

For instance, President Woodrow Wilson led America

through the horrors of the First World War – a slaughter in which nine million men were deployed against one another as human ammunition. More than a century later, the cold calculations that enabled generals to trade the lives of thousands of men and boys in a horrific bloodbath, for the sake of a few metres of French meadow, absolutely boggle the human mind. A famous moment early in the war, called the *"Christmas Truce"*, is a bittersweet reminder of the insanity of it. On Christmas Day 1914, boys on British and German sides emerged from their trenches to play friendly matches of soccer with each other before returning to their trenches, ready to continue the slaughter the following day. This momentary act of friendship reminds us that the brave working-class men and boys, mowed down in their millions on that war's battlefields, were there not because of any argument they had with one another. They were enlisted to kill or be killed because of a clash among European elites; because the political powers of the day considered this kind of slaughter of people to be the most expeditious way to settle a competition for territory. In short it was a reiteration of the same cold calculations as those of the Sky Council who sent our ancestors to war in the time of our earliest memory.

In the aftermath of that horrendous war, President Wilson issued the American people, and the people of the world, a dark warning concerning certain unaccountable powers, "Somebody... *something... a power somewhere so organized, so subtle, so watchful, so interlocked, so complete, so pervasive..."*

In the same vein, in 1961, having led America through the early Cold War, President Eisenhower issued a grave warning to humanity in his farewell speech to the nation. Sounding a disturbing note, Eisenhower warned of the shadowy influence of the *"Military Industrial Complex."* In that very same year President Kennedy recorded a speech which sounded a warning to the American people concerning the influence over governments exerted by *"secret societies."* All three presidential warnings hint at a fundamental lack of fellow feeling between humanity at large and the shadowy forces which can manipulate governments and foment wars. When I read these grim words against the backdrop of world mythology, I can't help but wonder: are these problems in human society even more entrenched than we think? Do they really go much further back than we may have ever imagined? Have our mythologies recorded the cynical workings of the Sky Council to help us understand not only how things were in the beginning, but the way things are today?

Looking at these dynamics raises a question for humanity at large, namely, what is our place in the great scheme of things? As you may recall, the Mayan story offers us a rather unflattering answer. At the very beginning of the human story, it says, those who engineered us accidentally produced a Homo Sapiens who was too smart to want to labour for others and too powerful to be easily corralled and managed. Our colonisers' chief genetic engineer, Quetzalcoatl *aka* Kukulkan *aka* Q'uq'umatz, had been tasked by his superiors, *"to make avatars for us to do the work and bring us our food."* The only intelligence required of these avatars was the intelligence to work effectively for their superiors. So, having overshot the mark, Quetzalcoatl was sent back to his lab to make a final adaptation to turn *Homo Sapiens Quetzalcoatlus* into us. This he did by dialling Homo Sapiens down to a level of consciousness in which our perceptual field is limited to the realm of our five natural senses. Hey presto! The perfect workforce.

It's a very cynical sounding story and no mistake! It brings no glory either to the Engineers or to their creations. It's hardly the kind of story a people would invent as their nation's glorious history. It also sounds some oddly familiar notes. When I first read the Mayan story it was easy to see how the argument among the Engineers of the *Popol Vuh* parallels the story of Genesis 3, when read with *"Elohim"* translated as *"Powerful*

Ones". It too reveals a conflict among those who engineered us over how intelligent we should be. I also hear a C21st iteration of this ancient conflict in contemporary conversations about higher education. Do we wish to raise students who can ask questions, challenge orthodoxies, think creatively, and see beyond the present in order to learn and make progress? In 2020, for instance, my friends in Australian education have been lamenting recent legislative changes, which appear to move the goal posts further away from those higher aspirations. There is a new watchword in the C21st – and it echoes in other countries. Our institutions must produce graduates who are *"industry-ready"*. As I sit at my desk with the Mayan text open before me, the language of *"industry-ready"* humans really bothers me. I can't help thinking that it has the timbre of that disturbing phrase in the *Popol Vuh*: *"... ready to be our avatars, to work for us and bring us our food!"* It makes me shudder.

But if the self-serving Engineers of the *Popol Vuh* sound cynical, it's no less disturbing than the account of things I grew up with in the pages of Genesis. In Genesis 3 the leading *Powerful One*, falsely translated as *"God"*, wanted Homo Sapiens to remain so unintelligent that we wouldn't even know we were naked. Evidently this *"God"* was no great champion of human education! In that light, why would human beings worship a superior who wanted such a low level of intelligence for us? Yet that is how the story goes. In the Biblical, Sumerian and Mayan accounts – the ET colonisers all realise the value of framing the people's subjection to them in the religious language of worship. They say in effect, *"If we can get the humans thinking that we who rule over them are better than they are, and are worthy of their worship – then it is so much easier to get the humans to work for us and bring us our food."*

The notion of human beings needing to worship or sacrifice to someone or something superior lies in the very bedrock of religious belief and practice. Through the historical dominance of religion in society, the principles of kowtowing to superiors has taken on the flavour of a virtue. That this should have happened within Christianity is all the more extraordinary given the various teachings that Jesus offered the world, intended to set people free from elitism, leader-ism and subjugation. After all, it was Jesus who said, *"Call no one on earth 'teacher'."* In other words, do not trust the pronouncements of any authority unquestioningly. He also said, *"Call no one on Earth 'leader'. You have only one leader and you are all brothers."* Hardly a ringing endorsement of monarchies, presidencies, or oligarchies – or indeed any other hierarchies! In another place Jesus talked about society's tendency to create pyramids of power and chains of human command with unchallengeable leaders at the top, and the people, pressed down and powerless at the bottom. He said simply, *"It must not be so among you."*

Equally importantly, the Jesus of the Gospels did not set himself up as a master to be worshipped or sacrificed to. For instance, when his followers implied that they would pray to Jesus after his death, he directed them instead to do as he did and engage directly for themselves with the Source as they would with their own Father. This is the practice that Jesus modelled. In another place Jesus told his followers emphatically, *"I did not come to be served."* Yet somehow Christianity has become oriented around *"serving Jesus"* or *"serving God"*. Indeed, the main business of many churches is the provision of *"services"* in which people can come and worship Jesus – a practice to be found nowhere in the teachings of Jesus himself. A generation later Jesus' famous follower the Apostle Paul mocked the very idea of human beings having to sacrifice to God, saying, *"The God who made the cosmos and everything in the cosmos... is not served by human hands – as if he had need of anything!"*

It's only logical. If we are using the word *"GOD"* as the Apostle Paul did to indicate the Cosmic Source, then any concept of sacrificial religion goes up in a puff of smoke.

However, from the perspective of cynical powers, there is a useful opportunity for leverage in a human population hard-wired to serve superiors.

Our most ancient stories survive for a reason. In the time I have spent with some of our oldest ancestral narratives I have come to an idea of why these stories survive and resurface with such remarkable resilience. In my years of training young pastors, from time to time, my most enthusiastic students would ask, *"What can I do to develop as a pastor?"* Of course, there are many possible answers – courses in psychology, listening skills, pastoral counselling and clinical therapy. Theological colleges generally offer a menu of such courses as a baseline. My thought is that to develop an understanding of how people tick, one of the best things you can do is read the literary classics. Why? Because if a story has lasted, it can only be because a succession of generations have connected with it. And there will be a reason for that. Clearly, people across a spectrum of times and places have found something relatable in the lives and events of those narratives. In other words, a story will endure if it tells us a recognizable truth about the world we live in. My own reading has led me to believe that there is, in reality, far less fiction in the world than we generally imagine. In fact, it is probably in fiction that we find the greatest freedom to tell the truth.

For instance, here is a story we all know. Retold and illustrated in countless children's books and anthologies, it was popularised by Hans Christian Andersen in the C19th. The story tells of some wicked merchants. (The audience boos.) One day the merchants in question decide to trick a vain King out of some cash. They offer to sell His Majesty a fabric which, they claim, only the very best people can see or feel. To uneducated or morally inferior people the fabric is virtually invisible. Naturally the King does not want to appear anything but the best and wisest, so he praises the merchants effusively for the fabric, which of course doesn't exist. When the King tries on a garment made from the invisible fabric, his courtiers wish to impress. Of course, none of the gathered courtiers wants to be exposed as uneducated or morally inferior, so they all heap praise on the King's new, non-existent garment with exaggerated *"oohs"* and *"aahs"*. Nobody wants to be the odd one out. On the occasion of his next royal procession through the city, the King decides to wear his new, invisible robes, which he does with great pomp and ceremony. That is, until a humble peasant boy, too innocent to stick to the script, simply believes his own eyes and blurts out the obvious: *"Look at the King! He's butt-naked!!"* In an instant the taboo is broken. The spell is lifted, and everything is laid bare – so to speak!

Versions of *The Emperor's New Clothes* can be found in the literature of C14th Spain and C13th India, each with a subtly different spin. The story's roots may go even further back, to the literary worlds of ancient Greece and ancient Persia. It has survived because, with a pithy humour, it offers the hearer a number of useful life lessons:

- *See with your own eyes and hear with your own ears.*
- *Have the confidence to trust your own judgement.*
- *Don't be silenced by a fear of power or a desire to cosy up to power.*
- *Don't be held back by a desire to fit in with the crowd.*
- *Beware of the influence of "wicked merchants" – people whose only agenda is money. They are always out to sell something – even if the thing being sold has nothing to it.*

The story of *The Emperor's New Clothes* is powerful because it makes these points through the drama and humour of a fiction so simple that even a young child can understand it, and it carries a special significance for anyone who identifies with beliefs or experiences outside of the mainstream. The pressure to see only what others in your group are seeing can be intense.

For instance, I can relate this to a friend of mine in pastoral ministry, whose parishioners have been troubled by various close encounters. When he sounded out his Bishop, he was told never to speak of these close encounters again. These phenomena fall outside the shared understanding of the clergy, so although his clergy colleagues are allowed to compare notes on a number of other paranormal phenomena, he has been advised to keep quiet about the disturbing thing his parishioners have been experiencing. Consequently, he will have to work out a pastoral response on his own. Of course, there is also a subtle hint of a warning that my friend may find himself looked down upon or cold-shouldered if he does speak up. It's the same dynamic.

Nobody wants to be looked down on as uneducated or inferior, so claims made by the academic consensus can be intimidating even to the most thoughtful and confident of scholars. I have been to the hieroglyphs of Egypt and have seen the carvings and sculptures of Central and South America. I have listened to the tortured debunking of some of ancient Egypt's most anomalous carvings and hieroglyphs, and of the ancient Central American carvings of space helmets and blue-tooths, and spacecraft. The academic consensus is that these carvings only look that way. The resemblance is coincidental and meaningless. In other words, *"Don't you see them that way, because the experts don't."* There's that same pressure again. It says, *"Better be persuaded or you will look stupid."*

The same imperative applies when it comes to not naming the obvious in various works of art in the Western tradition which appear to depict flying saucers and other anomalous technology. It is easy to be intimidated into not noticing or not naming these rather conspicuous artefacts.

For me, one of the most intriguing examples of an artistic anomaly is an article and woodprint illustration, published by a German broadsheet newspaper in 1561. The illustrated article reports an unusual event in the news of the day – a mass sighting of unidentified aerial phenomena over the city of Nuremberg. The illustration depicts a crashed object which has exploded, and an assortment of objects of various shapes navigating the air space over the city. The report, written up by a local printer, Hans Glaser, described the incident this way:

> *In the morning of April 14, 1561, at daybreak, between 4 and 5 a.m. a dreadful apparition occurred... and... was seen in Nuremberg in the city, before the gates and in the country – by many men and women... Two blood-red semi-circular arcs... a round ball of partly dull, partly black, ferrous colour... other balls in large number, about three in a line and four in a square, also some alone... a few blood-red crosses, between which there were blood-red strips... thicker to the rear and in the front malleable... among them two big rods... and within the small and big rods... more globes... These all started to fight among themselves... the globes flew back and forth among themselves and fought vehemently with each other for over an hour. And when... they became fatigued... they all... fell... down upon the earth "as if they all burned" and they then wasted away on the earth with immense smoke.*
>
> (Tr. Ilse Von Jacobi)

Interestingly Glaser encourages the reader not to speak of these phenomena *"with ridicule and discard them to the wind,"* not because he is proposing an ET explanation, but because he takes the events to be some kind of divine warning. The response Glaser advises is for the people to behave. *"Be good citizens!"* he says. *"Be good Christians!"* People should defer to civic and religious authorities, for their safety in such uncertain times. To give him credit though, Hans Glaser has done posterity a great favour by making a clear distinction between his own interpretation and what was actually seen. All this time later, what then do we do with what was seen?

In Bruges, Belgium you can see a famous tapestry, produced

in 1538 and titled *Summer's Triumph*. The anomaly within this work of art can be seen in the sky. Either the sky is peppered with flying cardinals' hats, or the objects depicted carry a more other-worldly significance. Another anomalous artwork can be found in the Church of the Dominican Monastery in Sighisoara, Romania. The church contains a painting which dates to the very early 1500s. It shows what appears to be a flying saucer, hovering over a building – possibly the monastery itself – with a beam of light, rising from it into the sky.

To be clear, the writers and painters I have just mentioned did not invoke the language of craft, airships or spaceships in their descriptions. Their messages are purely visual. By contrast the Eastern writers of the ancient Vedic stories had no such qualms. The ancient literature of the Vedas, curated by Hinduism, carry many stories of aerial battles among India's *"kings"* of ancient times. The Vedas describe craft called Vimanas which were part of the royal air forces. We are told they were mercury-fuelled vehicles with both aerial and space-faring capabilities. What is significant in these texts is that there is no obvious *"therefore behave"* that flows from the Vedic battle stories. And if these descriptive texts are not metaphorical or moral tales, then what exactly are we being shown?

While on Indian soil, we travel to the Raisen District of Madhya Pradesh. In the Phulsari rock shelters near the village of Sadlatpur, archaeologist Dr Wasim Khan has located an even older Indian reference to anomalous craft and ET phenomena. Etched into the wall of a cave is a glyph which Dr Khan dates at approximately 4,000 years old. It depicts a humanoid figure with a large head and large almond-shaped eyes. The figure is standing in front of a saucer-shaped craft, with what looks like a wormhole in the background. Now, someone might say that humanoids, flying saucers and wormholes are all concepts that were surely unknown to the ancient artist. That may or may not be so, but we can still place ourselves in front of these images, look with our own eyes and ask, *"What if the artists have simply drawn what they saw? And if so, what did they see?"*

In 2014 Indian archaeologist J.R. Baghat brought to the world's attention the rock paintings of Charama Chhattisgarh. They appear to depict a saucer-shaped craft with a tripod landing gear. Nearby stands a suited figure, sporting a helmet with antennae, and carrying some kind of probe. Perhaps the most striking element of the Charama Chhattisgarh rock paintings is the appearance of figures with elongated heads with narrow chins, wide crowns and large almond-shaped eyes. To a C21st eye it is impossible not to recognize the iconic features of small grey aliens. If this artist was simply painting what he saw, then there is really no question what we are looking at. And J.R. Baghat is not afraid to name the obvious. He says, *"The findings suggest that human beings in prehistoric times may have seen, or imagined, beings from other planets which still provoke curiosity among people, and researchers."*

What were these ancient artists commemorating for posterity? Were they memorialising something imagined, or something that happened?

Some equally intriguing etchings can be found in caves at Tassili N'Ajjer in Algeria's portion of the Sahara. They depict a figure in what resembles a helmeted spacesuit. In the background a number of saucer-shaped craft perform manoeuvres, accompanied by smaller orb-like objects. Again, what does it tell us if the artists simply etched what they saw?

At the National Museum of Guatemala, you can visit an entire section devoted to carvings of figures wearing what the modern eye cannot help seeing as helmets and blue-tooths. The conventional explanation is that this is what ancient priests wore for their commemorations and ceremonies. But to my mind this explanation automatically invites another question: *"Why?"* Why does their ceremonial headgear look that way? Who were the wearers trying to resemble? And why did those particular motifs come to symbolise advanced power and higher intelligence?

If I allow myself to see with my own eyes and trust my own judgement, then I can perceive that the world's artistic canon has some intriguing stories to tell – stories which may differ to our canon of written texts and official histories.

For another example, we can look at *The Annunciation*, a 1648 painting by Italian renaissance artist Carlo Crivelli. The painting depicts an event from Luke's Gospel, the supernatural conception of Jesus. In the picture the Virgin Mary is shown wearing some unusual headgear on her forehead. Focussed directly on to a crystal in her headgear is a laser beam, projected by a disc-shaped object hovering high in the sky. The object bears a striking resemblance to the craft depicted in the painting found in the Church of the Dominican Monastery in Sighisoara, Romania, from the previous century. Suffice it to say, Crivelli's portrayal is nothing like the written account of Luke's Gospel. What then is the artist depicting? And where has this idea come from? Is it a random creative flourish? Or has Crivelli left in plain sight a message for any open-eyed viewer? Is it an alternative explanation of the conception of Jesus?

A similar motif can be found in a work of art from 1710 by the Dutch artist Aert de Gelder. His painting records another significant moment in the story of Jesus. It shows Jesus being baptised by his cousin John. Brooding over the scene, high in the sky, is an enormous disc-shaped object which is focussing four laser beams on to the bodies of John and Jesus below. Once again, the artist appears to be proposing something other than the familiar Biblical version of the story. Is de Gelder offering the viewer an alternative narrative concerning Jesus' special abilities?

Of course, the sceptic will say, *"That's not a spacesuit. It just looks like one,"* or *"Educated people understand that's really just a moral tale,"* or *"Those who are qualified to comment on this painting understand that this is simply how supernatural events are traditionally depicted,"* etc. Once again, we are in the realm of *The Emperor's New Clothes.* Speaking personally, the first time I saw each one of the anomalous carvings and paintings I have just described to you, I actually thought they were spoofs. In reality, all the artefacts I have just mentioned are in the public domain for you to see with your own eyes and interpret for yourself.

At this point, you might be wondering how artists like Carlo Crivelli or Aert de Gelder could possibly be the curators of secret or privileged information. Why would an artist have access to a different world of knowledge to the rest of us? To answer that question, think again about those moments in history when, through an invasion, or a coup or through new legislation, governments have made themselves the arbiters of what is news and what is fake news; what is information and what is disinformation. Think about the moment when Emperor Theodosius added imperial force to Christian orthodoxy and sounded an intimidating warning to any group wishing to argue for a different view. Think about that moment when the Catholic invasion of Central and South America led to the execution of the indigenous priesthoods, and the burning or burial of their sacred texts. In such moments of suppression, the banned body of knowledge gets despatched to two places. Firstly, it gets archived in the libraries and vaults of emperors and popes and so becomes the preserve of political and religious elites. Secondly, it gets hidden and buried for its own protection, for example, the Gnostic Gospels, buried in the caves of the Nag Hammadi desert. This means that the second group with access to the forbidden knowledge comprises the underground priesthoods and the secret societies who know where the treasure is buried.

It is often from this second locus that the world's forbidden knowledge leaks into the world's artistic canon. Sometimes the artists in question are themselves the members of such esoteric societies. Other times artists may be offered access to artefacts or information to be worked into a painting, a carving, a novel or a movie-script. In this way forbidden ideas, such as those of the Gnostic Gospels, continue to surface and resurface long after their banning, burning and burial. There is always someone somewhere keeping the forbidden knowledge alive and leaking it, one idea at a time, back into the public sphere. This is how esoteric messages get passed down through the ages for the eagle-eyed to spot.

Just when I think I have got my head around this strange world of encoded art and hidden messages, a conversation with a renowned scientist in Kazakhstan takes the story of secret messages from the past into a whole other dimension. In the next chapter this senior researcher in astrobiology and astrophysics will point us to C21st research indicating that esoteric messages have been passed to us not merely from previous generations, but from previous civilizations; from people on other planets, orbiting other stars.

Chapter Ten

We Have Been Here Before

Kazakhstan – 2013

"It is not about aliens!"

Maxim Makukov is negotiating the buzz of media attention and spin following the publication of his cutting-edge research in the science journal *Icarus*. The research in question has been the collaboration of Maxim and Vladimir shCherbak of the Fesenkov Astrophysical Institute and the al-Farabi Kazakh National University of Kazakhstan. Kazakhstan is an enormous country. In fact, it is the largest landlocked country in the world. Yet it is not often on the front pages of the Western Press. However, Makukov and shCherbak's research has it in the headlines and the implications of what they have found are ground-breaking.

To give you an idea, the 1997 movie *Contact* – written as a novel by astrophysicist Carl Sagan – was based on the idea that a benevolent neighbouring civilization might send a signal to planet Earth, embedded with non-random mathematical patterns as a way of cluing the receivers that the source of the broadcast is an extraterrestrial intelligence. Makukov and shCherbak argue that their findings demonstrate beyond reasonable doubt that just such a signal has already been sent. It's embedded in our genetic code. Coming only four years after the Vatican's invitation to embrace a *"brother or sister alien,"* the newspapers are all over it. Some leap upon Makukov and shCherbak's paper as proof of God and creationism. Others claim it as proof of aliens and evolution. To my surprise the media are positive, even if with a raised eyebrow or two, as they applaud the courage of these two Kazakh scientists, who apparently have spent thirteen years studying the human genome. This last detail, however, turns out not to be true.

Canberra, Australia
August 2020

Makukov and shCherbak did not work on the human genome. Not even for a year, let alone thirteen years, as I had read in countless newspaper reports at the time. This is why in 2020 I have taken my questions directly to the source, cap in hand, to sit at the feet of Maxim Makukov himself and listen first-hand to what he and his collaborator Vladimir shCherbak actually intended in all that flurry of excitement now seven years ago.

"The 'signal' sits in the genetic code," Maxim explains. *"Which is something very different from the genome."* This means that all life on Earth would owe a debt to the seeders, who Maxim identifies very precisely as a parent *"civilization... an ET intelligence"* – not aliens!

"The very word 'aliens' is a misnomer," he says, *"since the senders and the seeds' descendants share common cellular ancestry."*

My ears prick up when I hear this. Maxim's very precise language echoes of the points made by Reverend Doctor Guy Consolmagno at the time of the Colloquium back in 2009. He too emphasised that we should not use the term *"aliens"* when talking about our cosmic neighbours. After all, he said, we would be creatures of the same creation, children of the same heavenly Father. Maxim is making the same point, though not on the basis of theology. He is speaking purely about what he and his collaborator observed in the mathematical patterns of our genetic coding. He even has an idea of what mechanism might have been used to spread the seeds of life through our galaxy. *"Seeding star-forming proto-clusters,"* he says, *"might be more efficient than seeding individual planets. If that was the case, it is likely that the stars which were born in the same cluster the Sun was born in might host life as well."*

In my mind, this conjures up a vision of the *Star Trek* universe in which the United Federation of Planets brings together a range of species who, for the most part, look suspiciously similar to one another: upright, two legs, two arms, a neck, and a head with a pair of eyes, a pair of ears, a mouth and pair of nostrils. I have always assumed this similarity was a function of budgetary constraints for costumes and make-up at the Desilu Studio. What Maxim is saying hints at a real world scenario.

"So, Maxim, if at a cellular level we're related to all the life-forms on planets orbiting related stars, how recognizable would we expect our genetic relatives to be?"

"That's a tough question," he answers, *"[if] they had evolved to some higher forms at all. Currently there is no answer, only intuitive guesses. Still, I would note that many evolutionary biologists now consider 'Convergent Evolution' as a major player. Under this view it becomes likely that our intelligent 'cosmic relatives' might not be very different to us – as per appearance."*

Convergent evolution is the theory that the same genetic coding will lead ultimately to similar creatures for similar niches. As a citizen of Australia, I see the outcome of convergent evolution around me all the time. For instance, Australia has no groundhogs. We have wombats, the more excellent marsupial version. Groundhogs and Wombats are not related. They have just developed a somewhat similar form for a similar niche. In Australia we have no indigenous dogs. But until 1938 what we did have was the Tasmanian Tiger which, to the untrained eye, was a dog-sized carnivore indistinguishable from a dog. Except it wasn't. It was a marsupial. For another example we have Tasmanian Devils in the place of wolverines, echidnas in the place of porcupines, and instead of flying squirrels we have sugar gliders. I could go on. In a way Australia provides a good Earth-based illustration of what might happen if the same genetic seeding that landed on planet Earth were to land on a planet with similar conditions. For that reason perhaps we should not be surprised if interstellar neighbours should turn up looking rather human.

Essentially, what Makukov and shCherbak's findings have done is put flesh on the bones of a theory which was championed by Francis Crick, the Nobel prize-winning co-discoverer of the double helix of DNA. Among other eminent scientists and DNA researchers, Francis Crick argued that all life on Earth was ultimately of extraterrestrial origin, and that the genetic code for conscious, intelligent life was deliberately seeded throughout at least this part of the cosmos. *"I am rather sceptical,"* Maxim says, *"about the possibility of ET intervention into terrestrial evolution, all those parallels in ancient scripts notwithstanding. I do not say this is ruled out completely."*

If I understand Maxim correctly, the purpose of the non-random code sent out to seed the cosmos was to ensure a continuation and spreading of not merely life, but specifically life with consciousness and intelligence – what Maxim describes as *"intelligence plus experience."*

"[It] adds to my experience," he says, *"[the] knowledge that there existed a technology-oriented culture elsewhere before even Earth was formed."*

Maxim's language is emphatically precise, carefully keeping clear blue water between objective findings and any personal speculation. But I am eager to see if I can draw another *"therefore"* from our conversation.

"Maxim, if an ET intelligence engineered the code that seeded planet Earth, does that mean that our genetic coding is showing us a snapshot in a chain of old civilizations seeding new civilizations? Have you found anything that would indicate a zero point or an ultimate source, given that our universe appears to have a finite beginning?"

"No," he says. *"Nothing we found indicates a chain, although... it also does not preclude it. My personal educated guess, based on available astrophysical data, is that we are the second in the series. It is largely more probable that our direct ancestors were the zero point,*

rather than they also had descended from seeding."

I am grateful to Maxim for his courage and precision, making a mental note that the arrival of this genetic coding on Earth may have seeded terrestrial civilizations long before anything we know about. As for what we do know about terrestrial civilization, most school textbooks point to the Epic of Gilgamesh as the oldest narrative of the Earth's oldest known civilization. However, it is clear from the very text itself that the Epic of Gilgamesh is quoting a story from a source even more ancient than itself.

Southeast Turkey is one place that hints at the possibility of a relay of successive civilizations on planet Earth. In 1998 the University of Norway of Applied Sciences and the Max Planck Institute of Cologne, Germany sent a team led by Professor Manfred Heun to the mountain region of Karaca Dag in Southeast Turkey. There they explored the surviving evidence of a world-changing technology from the time of the last ice age. In the tipping point of human history, as the ice sheets receded, one tribe, or even one family, as Manfred Heun suggested, made a discovery that was to be the seed of our civilization. They learned to farm. Farms produce surpluses. Surpluses produce social specialisation. From this flows the ability to develop complex, stable societies, and to build cities to house them. It was an incredible leap forward.

To give you a sense of scale, let me tell you that I used to live on the farm in Southeast Australia where William Farrer famously created Federation Wheat. At the end of the nineteenth century, urban Australia desperately needed a form of wheat which could reliably generate large crops, season after season, in the harsh conditions of the Australian climate. William Farrar was a scientist and the son of a farmer. With a lifelong knowledge of farming and all the advantages of C19th science, Farrar was able to genetically modify strains of wheat until he had a plant which could be cultivated as a crop in Australian soil. To make that modification it took him twenty years. The *"family"* of 10,000 years ago, identified by Manfred Heun's research, had to do what Farrer did eleven times over to turn the naturally occurring plants they adapted into cultivatable crops. Somehow the agricultural science pioneered by this enigmatic family then spread – which it did rapidly – all around the planet. Clearly, we must be looking at a very talented family because, as well as working out how to genetically modify those eleven crops, they also learned the art of animal husbandry. Manfred Heun's team has given the world a lot to think about.

Curiously, the remnants of an even earlier epoch can be found just a few hours down the road from Karaca Dag. Gobekli Tepe is a sophisticated megalithic site, perhaps fifty times the size of Stonehenge, once fully excavated. Carvings in the rock hint at astronomical expertise and international connections with other ancient sites. The structures are estimated at 10,000 years of age. Even more curiously the site at Gobekli Tepe appears to have been very carefully buried as a means of preservation around 8,000 years ago. Don't miss those dates. In locations within hours of each other in ancient Southeast Turkey it appears that we are witnessing the end of one megalithic culture and the beginnings of another. The proximity of time and place makes me wonder if the family at Karaca Dag may have had a little assistance from their forebears at Gobekli Tepe in the process of seeding today's civilization.

The oral history of the Mohican people of the upper Hudson River Valley tells of exactly this kind of outside assistance. Their story speaks of a time long ago when some kind of catastrophe had forced their ancestors into higher, less hospitable terrain. In this harsh and unfamiliar territory their ability to survive was suddenly uncertain. The Mohican people credit their survival to the sudden appearance of certain mysterious others, who coached them in technologies which would enable them to weather the change, feed and medicate themselves, and make it through for a more prosperous future. I wonder if our friends at Karaca Dag may have benefitted from a similar intervention when their own survival hung in the balance.

While a relay of terrestrial civilizations might not be too hard to imagine, some may find it a stretch to accommodate Makukov and shCherbak's model of an interstellar relay. Yet that idea too resides in our aboriginal stories. Whether sitting at the feet of the Dogon people of Mali, West Africa, listening to their stories of the Sirius star system, or sitting at the feet of the Efik people of Nigeria as they tell their story of Abassi and Atai, we are soaking in a story of interstellar seeding which goes back thousands of years. The Zulu people also have their story of life arriving on Earth from outer space. Unkulunkulu, the first human, lands on the planet's surface, along with every living creature, in seed form. The new arrivals develop in seed pods on Earth's fertile soil, until ready to burst out on to the planet's surface. It is a wonderfully cinematic version of panspermia. Clearly the idea of an intergalactic parent civilization is not new.

If we can accept it, we are not the first civilization, neither in space nor on our own planet. We are perhaps only the latest of many. My reading of Genesis suggests that by the time we reach the end of the stories of beginnings (*Genesis 1-11*) we have the recollection of possibly three different resets of civilization on a global scale.

- In *Genesis 1* and 2 the planet is terraformed following a catastrophe which has flooded the planet and shrouded it in darkness.
- *Genesis 6* tells the story of a flood of extinction level proportions. Those who populated the planet prior to the flood lived lives many times the lifespan of the human species that followed post-flood. This is a detail repeated from the Sumerian accounts.

- *Genesis 11* speaks of an attack on a technological human civilization so profound that the survivors had to re-evolve the faculty of speech.

It is possible that these three narratives offer different recollections of the same planetary reboot. Or they could be the memories of three different cataclysms. Whichever it is, some traumatic memory has been scarred into the fabric of Genesis to show that we are not the first civilization on Earth, and we have known it for a long time. Neither are we the first technological people to have graced this planet. Reading Genesis and the Sumerian cuneiforms alongside one another reveals that the technology attacked in the Babel narrative is space-faring technology. Etymologically the word *"Babel"* means *"gateway for the Powerful Ones"*. The Genesis account describes the function of Babel's towering structure as a bridge to the heavens. The Sumerian texts tell us that from this structure fifty technicians despatch three hundred *"observers"* to their stations in the stars – what we would call *"space stations"*. These narratives, alongside Indian Vedas and carvings around the world, tell us that Earth has been home to space-farers before. Now, in the C21st, images of anomalous artefacts on the surface of Mars are raising the possibility that when we finally land a manned mission on its surface we will not be the first civilization to tread the soil of our planetary neighbour. If we take our world mythologies as vehicles of ancient memory, such a find on the Red Planet need not surprise us.

Back on earth, archaeological finds continue to reveal an unacknowledged past on our own planet. Off the coasts of India, Japan, Malta and Cuba, under up to thirty-six metres of ocean lie the megalithic remains of cities which would have been above sea level no more recently than ten thousand years ago. Yet somehow our school textbooks still report that civilization began no more than seven and half thousand years ago in the fertile crescent. Today, archaeological voices are increasingly telling us there was a civilization before. New technologies such as ground penetrating radar, and soil magnetization analysis are revealing the presence of forgotten cities buried beneath the jungles of Cambodia and Amazonia. These are not prehistoric cities. Indeed, their zenith was during the common era. Yet until the last few years nature has kept them hidden them from sight, demonstrating just how quickly the signs of megalithic civilizations can disappear from view. For instance, when Machu Picchu in the Peruvian Andes was photographed in 1915, not much more could be seen than a mass of jungle. Seventy years later excavations revealed the beautiful megalithic remains of the ancient Inca, whose culture spread across the Andes of South America and into Central America.

In 1977 the Dutch anthropologist Maria Scholten published her work, *La Ruta de Viracocha*. Her work demonstrated the sophistication of the megalithic cultures of the Andes and pointed to evidence suggesting an even older *Mother Culture*, which predated and connected them all. The cities of the Inca are fascinating in their own right. The design of Tiahuanaco speaks of advanced city planning in alignment with the stars and the phenomenal fluid engineering of stone at Sacsayhuaman at Cusco has become a byword for mystifying and awe-inspiring skills lost to today's civil engineers. Maria Scholten's work demonstrated that the major Incan cities of Tiahuanaco, Pukara, Cusco, Ollantayambo, Machu Picchu, and Cajamarca were constructed at precise intervals along a North-West to South-East vector, which sits on an exact bearing of 45 degrees from the North-South axis. The ability of ancient city-builders to coordinate construction so accurately between sites across a 1,500km vector, through the Andean mountains, in what today are separate countries, is absolutely mind-boggling. Furthermore, the axis – called the *Ruta de Viracocha* – not only joins the cities themselves but runs precisely through the diagonals of the major temples of each city, perfectly intersecting them.

Maria Scholten was also able to identify the unit of measurement employed by this forgotten Andean *Mother Culture*. The *"Andean Unit"* was 3.34 times the units of today's metric system. The unit for cloth and ceramics was 3.34cm. For buildings it was 3.34m. For distance it was 3.34km. The significance of this correlation is that it confirms for us that the Andean *Mother Culture* based its system of measurement on the size of the planet – something that our civilization didn't do until 1795. The *Andean Unit* was based on one-thirtieth of a millionth of the distance from the equator to the pole. Evidently, the Andean Mother Culture knew how to measure the size of the planet. Furthermore, the alignment of the cities and structures throughout the Ruta de Viracocha demonstrates precise alignments with the stars of the Milky Way. This would suggest that the highly developed astronomical knowledge for which the Incan people are renowned actually goes back to the Andean Mother Culture – the civilization that came before.

The descendants of the Inca of Cusco tell a story about the origins of the Mother Culture. Their story was relayed to Spanish linguist, Juan Diez de Betanzos, in the 1550s. His *Narrative of the Incas* described a powerful being, called Viracocha, whose intervention in the story of planet Earth parallels many of the elements of the *Popol Vuh* and the stories of Kukulkan/ Q'uq'umatz/Quetzalcoatl. Viracocha also emerges in that mysterious epoch when planet Earth is shrouded in darkness. He too dislikes his early attempts at generating a suitable Homo Sapiens. Just like the Engineers of the Mayan story, Viracocha genocides his failed attempts at humanity with a devastating flood.

The *Narrative of the Incas* goes further, though, and credits Viracocha not only with the final engineering of Homo Sapiens but also with humanity's basic education in agriculture, medicine, engineering, and all the basics of city-based civilization. At the same time, certain aspects of the Incan account as reported by de Betanzos appear to have fused the figure of Viracocha, who first appears over the waters of Lake Titicaca, with the cosmic figure of Saturn, thus associating Viracocha with a powerful, cosmic, deity figure.

In 1907 another Spanish scholar, Pedro Sarmiento de Gamboa, relayed the stories told to him by the people of Cusco. According to de Gamboa's version of events, Viracocha – although known as *"The Creator"* – arrived on an already existing South America, albeit one shrouded in darkness and in need of light and terraforming. Significantly, Viracocha was described in this account as being a man, of medium height, with light skin. He carried a book of wisdom, wore white garments, and walked with a staff. Not quite the cosmic, god-like figure one might have expected. Sadly, this indigenous recollection of Viracocha's appearance may have left Incan descendants rather vulnerable to the claims of the Catholic forces who invaded in the C15th and C16th. The European invaders brought a book of sacred wisdom, and they were endowed with light skin. Their officers dressed in white garments and carried staffs which fired bullets. The correlations were so obvious that when Francisco Pizarro landed on the Peruvian coast at San Mateo Bay in 1532 the locals took the Conquistador to be Viracocha himself, finally returned – a fortunate (for him) confusion. The Catholic educators quickly decided to reaffirm the well-established mythology of Viracocha and assimilate it to the Christian story as a way of annexing the people's existing religious loyalties. It was easy to recognize similarities between the Andean stories of Viracocha with the Bible's stories of beginnings. The Biblical story could piggyback quite nicely on a story the locals already believed. It was a convenient way to fast-track the people's evangelization to assert that Viracocha was God.

The theological sleight of hand shown in this historic moment provides us with a powerful object lesson. In exactly the same way that the Bible's redactors equated God with the *elohim* and so credited God with all the violence and brutality of the *Powerful Ones*, so the Spanish missionaries of the sixteenth century equated God with Viracocha and in so doing credited the Christian God with the genocide of the people's ancestors, memorialised in the Incan narratives. This confusion anchored the idea of a violent God who could not be questioned – an image helpful to invaders sporting both Bibles and guns. At the same time, the equation of Viracocha with God obscured the pointy question of precisely what kind of entity Viracocha really was. The effects would be far-reaching. Think about what happens to the heart and mind of a child brought up in the home of an abusive, violent, alcoholic parent, a parent who has to be appeased and tiptoed around. It eviscerates the child's happiness and self-esteem. To the extent that we have bought into a universe that equates GOD with a violent, genociding entity who cannot be questioned we have placed the whole of humanity into such a household. You have to wonder what that equation has done to the psychology of human beings. This is why confusing GOD with other entities, whether Viracocha, the Elohim, the Sky People or the Engineers, is such a critical error.

If we look back to the history that preceded this unfortunate moment of confusion, it is fascinating to see that the Inca people clearly affirmed an external intervention by a non-human entity as their way of accounting for their ancestors' intelligence and facility with the basic technologies of society. Just as the proximity of Gobekli Tepe to Karaca Dag within Southeast Turkey hints at the possibility of a previous culture assisting a new one, so the ancient peoples of the Andes attest to the advancement of a previous civilization with roots off-planet.

Plato argued that Earth has been home to a succession of civilizations. From Egyptian sources he became convinced that the movement of objects in space impact planet Earth on average once every five thousand years, each time taking civilization on

Earth down to a virtual zero. The testimonies of the Andes, of Southeast Turkey and the lost cities of our coastal waters would all suggest that Plato was on to something.

What if the story of planetary reboots goes even further back into geological timescales? If civilizations preceded us by geological timeframes how could we possibly know? *Genesis 11* hints at a time when the Earth had a single landmass and a single coastline. *Genesis 10* names the generation of Peleg as the time when Earth's continents first separated. Similarly, there is an Australian Aboriginal story of the separation of Australia and India. By today's reckoning the timeline of Homo Sapiens extends only 200,000 years into our history – well outside the timeframe of such geological adventures. However, it is not impossible to imagine that civilization on Earth may have a longer and more interesting story than anything we have been able to dig up. After all, if there were a technological species thriving on planet Earth long before us, long before the dinosaurs, before the Cambrian explosion, by now all the material evidence of such a society would be compressed within a few millimetres of minerals in the rocky sediments of our planet. Absent of another source of information, we simply wouldn't know.

As to the story of our own civilization, the many conflicts among empires competing for territory and the warfare between colonisers and indigenous peoples have all left deep scars in the soul of our current civilization. Indeed, the competition of cultures is as alive today as ever. What lies at the heart of these conflicts?

The oldest continual culture on planet Earth is that of Australia's Aboriginal peoples. Their closest relatives, according to the latest DNA research, are the Native Americans of South, Central and North America. And there are fascinating cultural parallels in their respective ways of living in harmony with the land. Like their Native American cousins, Aboriginal

Australians speak of visitors from the Pleiades, and credit them with their ancestors' tuition. The Aboriginal presence on Australian soil is understood to go back more than 60,000 years. So, theirs is surely our supreme story of living sustainably on this planet. The intervention that created farming in Southeast Turkey 10,000 years ago has a different aspect. It begins with the modification of genes in a naturally occurring grass to turn it into wheat for cultivation on a massive scale. Karaca Dag initiated a model of farming which would lead to specialised society, city-living, and ultimately, to the industrialisation of food and farming.

The story of Australia dramatizes the difference between these two agricultural interventions. British Australia needed William Farrer's Federation Wheat – a genetic adaption in the tradition of Karaca Dag – in order to support cities on a European scale. Meanwhile Aboriginal Australians had been living on the land quite comfortably and sustainably, using Australia's naturally occurring grasses to make their flour and bread for more than 60,000 years. The two different wheats illustrate two different approaches, two different ways of living on the continent.

The location of Karaca Dag in Southeast Turkey hints at the possibility that its lesson in agronomy of 10,000 years ago may have come from the same sources who the Babylonian mythology says provided our Mesopotamian ancestors with their lessons in banking, money and legal systems, contracts, time-measurement and record keeping. It makes me wonder about today's battles between industrial scale, GM, petrochemical based farming (typified by corporations like Monsanto) and the traditional approaches of farmers around the world who use natural organic seed, and the natural synergies of combination and rotational farming. Are those battles, in reality, the inevitable outcome of two conflicting models of farming, imparted through two quite different interventions in our evolution as a civilization – the most recent, 10,000 years ago, just after the most recent planetary catastrophe, and the other after the previous cataclysm, more than 60,000 years ago? Are we today caught in a competition of cultures with origins far beyond the world of our Earthly ancestors?

If our world mythologies genuinely carry an ancestral memory of prehistoric tutors from the stars, I wonder how those ancient visitors must have appeared to have gained such traction with a primitive human population who had never seen them before? What was it about our visitors that would have inspired respect rather than repulsion, admiration rather than pure terror? To answer that question, our journey will take us first to the Levant, in the early iron age, and from there to an unsettling close encounter in the English city of Chichester.

Chapter Eleven

Meet the Sky People

The West Bank – 1250 BCE

We are on high ground in the shelter of a wooded ridge, looking out across the fertile lands of the Levant. It is populated by people who are so big that the spies returning from reconnaissance are questioning whether an invasion is even possible. Their report is unnerving, *"They made us feel like grasshoppers."* Why then are these people so gigantic? They are the *"Anakim"* – a people-group whose history goes all the way back to the beginning times when *Benei Elohim* took human females and produced the giants we call *Nephilim*. The *Anakim* have been among us ever since.

If you're feeling a bit disoriented, we are in the book of Joshua, hearing the writer reaffirm the Hebrew memory of a hybridization which combined the traits of humans with *"The Ones like the Powerful Ones"* (the *benei elohim*). The fact that this hybridized ancestry is memorialised with the name *Anakim* is striking. The word *Anakim* is uncannily close to the Sumerian name for our ancient hybridizers – the *Anunnaki*. According to the Hebrew tradition, *Anak* is the progenitor of the giant *Anakim*. In Islam, *Anak* has a female counterpart called *Anaq*. She is the mother of the giant king *Uj*. In the Hebrew texts *Uj* is known as *"Og"* and he was the governor of the people of Bashan. Both religious traditions claim he was a giant. In fact, Og aka Uj was so gigantic, so the Hebrew story goes, that his bed was a full six feet wide by thirteen feet long. In a similar vein, Greek culture memorialises an ancient ruler by the name of *Anax* – who governed over the giant people known as *Anactorians*. They, apparently, were fifteen feet tall.

The association of such similar names with the mythology

of giant people is one of those tantalising clues suggesting that real people groups with real names are being recalled in our ancestral memory.

Stories of giants crop up in mythologies all around the world. Celtic, Norse, Hindu, Native American and Aboriginal Australian traditions all speak of giants sharing the planet with our distant ancestors. If these stories are rooted in ancestral memory it would be reasonable to expect some giant remains to have been found – and tested. As it happens, finds of anomalously large human remains have indeed been made – and most frequently in the Americas. In the USA, between 1819 and 1959 finds of human remains, measuring between seven and twelve feet tall, were unearthed in Ohio, West Virginia, Iowa, Missouri, California, and Pennsylvania. The finds were reported in the press, and for a time some of the remains were curated by the Smithsonian Institute in Washington DC. Extra fingers, jaws full of molar teeth, and other anatomical anomalies besides, all cry out for analysis and DNA testing to establish who our North American neighbours might have been. However, the question remains unresolved to this day because apparently all the giant remains have now been repatriated in accordance with the Native American Graves Protection and Repatriation Act. The repatriation of these bodily remains is certainly a respectful act but, absent of DNA testing, we are left to wonder if these tall beings were part of a different genus altogether or simply very tall and unusual people.

The same tantalising question applies to another group of mysterious visitors to the Levant in the second millennium BCE. The circumstances of this encounter – described in Genesis 18 – relate to the conception of a child named Isaac, the son of Abraham and Sarah – the progenitors of the Hebrew tradition. Abraham and Sarah emerged from a culture with Sumerian roots. They probably spoke Akkadian, the lingua franca of the daughter cultures of Sumeria. They would have grown up schooled in the Sumerian stories of beginnings – the ones about the *Sky People*. As we have already seen, Abraham and Sarah's tradition had its own name for the *Sky People*. It called them *"The Powerful Ones"*.

The presence in *Genesis 18* of the holy name YHWH clues the reader straight away that the Abrahamic story has been retold by a writer from after the time of Moses. We can be confident that the word that YHWH has pasted itself over in the original telling is the word *"elohim"*. The importance of this moment in *Genesis 18* is that Abraham and Sarah meet three of them, in the flesh. It is an encounter that reveals to us what *Powerful Ones/ Sky People* actually looked like, face to face. The story begins like this:

*Some Powerful Ones appeared to Abraham near the great tree of Mamre, while he was sitting at the entrance to his tent in the heat of the day. Abraham looked up and saw **three men** standing nearby.*

Evidently, what Abraham can see is three men. No other descriptor is given at this stage, no anomalous size or shape. They just look like men. Abraham responds to his three visitors with the kind of warm hospitality that is one of the beautiful things embodied by the cultures of the Levant. He addresses his three guests as *"Sirs"* and he and his wife then prepare some food and allow them to eat. Following the meal, the conversation takes a rather unusual turn when the visitors explain to Abraham that they will be passing this way again in exactly one year, by which time, they say, he and Sarah will have produced a child. On hearing this Sarah can't help herself laughing and, equally nonplussed, Abraham plies the visitors with questions, since he and his wife are both well past childbearing. Who or what are these *"men"* to make such an impossible claim? And, by the way, the reader knows that Abraham and Sarah will indeed produce a child within the year.

What is significant is that throughout the encounter with the three *Elohim/Sky People* all Abraham and Sarah saw were three men. Apparently, that's what *Sky People* look like. They appear human. After their departure though Abraham and Sarah must surely have asked each other, *"What just happened?"* And perhaps it was only after realising that Sarah had indeed fallen pregnant that they wondered at the true nature of their strange visitors.

Following their enigmatic visit to Abraham and Sarah, the three *Sky People* then continue their journey toward the ill-fated city of Sodom, to visit Abraham's relatives. In the account that follows the text refers to the men as *"angels"*, although it's worth noting that the word *"angel"* really means no more than an *"agent"* – a person sent with an assignment or a message. It doesn't denote anything about the agent's biology or genus. Again, the editor of the Sodom story refers to the visiting *Sky People* simply as *"men"*. In this passage the editor has reduced their number to two. I wonder if this may be to separate the holy name of YHWH from the embarrassment of what is about to happen. He also tells us a little more about the appearance of the *Sky People* as they arrive in Sodom. They are incredibly attractive. Unfortunately, in Sodom, this becomes a problem. When a violent gang mobs the house where the *Sky People* are staying and threatens them with sexual assault they escape, taking Abraham's relatives with them, and demonstrating something else that marks *Sky People* out from the average human. They blitz the city behind them with a powerful destructive technology that is like nothing the world of Abraham had ever seen. These dramatic moments play a pivotal role in the longer narrative of Abraham and the Abrahamic peoples. But if we can accept these two *elohim* narratives as vehicles of some ancient memory then we have allowed ourselves a physical glimpse of the *Powerful Ones/Sky People*. Apparently, they look quite human, and are very attractive.

More than one and a half millennia later, the Gospel of Luke tells the story of a Jewish Priest, a descendant of Abraham and

Sarah, named Zechariah as he burns incense in the Jerusalem Temple. While worshippers are gathering outside the sanctuary area, an unusual messenger appears in the sanctuary. Something about the way the messenger appears in the room startles and terrifies the elderly priest. *"Don't be afraid,"* the messenger says. *"Your prayer has been heard. Your wife Elizabeth is going to give birth to a son. Call him John and follow the instructions I have for you. Your son will have a special assignment."*

John's is another anomalous birth in the Judaeo-Christian tradition. John's parents Zechariah and Elizabeth, just like their ancestors Abraham and Sarah, are well past the age of childbearing. And John, of course, did prove to have a special assignment. As the famed John-the-Baptist, his prophetic voice in the religious and political world of first century Judea carried an importance that would be hard to exaggerate. In fact, the story surrounding John's birth takes up more space in the Gospel of Luke than the birth of Jesus. And John's audience appeal extended far beyond the Jewish religious world of his day. Indeed John the Baptist's reputation was so great that when his cousin Jesus of Nazareth began his own preaching tours many believed Jesus to be a reincarnation of John, and after the Baptist's unfortunate execution a great many fully believed that John would resurrect. Jesus held his cousin John in the highest regard, describing him as the greatest man in history. That's quite an accolade coming from Jesus Christ!

Of course, we might expect a grand entrance for such a historic figure. But I find it curious that the birth of John-the-Baptist is announced to his father through a close encounter so similar to the story of Abraham's enigmatic visit from the *Sky People*. According to Luke's Gospel, the same unusual messenger then visits Elizabeth's cousin Mary to tell her that she also will be giving birth to a son. Once again, it is the story of a conception that is achieved artificially, announced in a close encounter and, just as in the story of Elizabeth and

Sarah, no sexual intercourse is mentioned. Of course, it is entirely possible that Luke has simply made these stories up. It is striking that there is no parallel to them in the other three canonical Gospels. If they are Luke's fictional invention, then it appears that he has based them on a close encounter with *Sky People*. The alternative possibility is that both Elizabeth and Mary genuinely experienced their own close encounters in association with their respective pregnancies.

The story of the hybridization of humans by Sky People is arguably the oldest and most widely recurring theme in the world's ancestral narratives. It has continued from ancient times until today. The family of Mami Wata narratives among African peoples around the world and the fay tradition of Europe's Celtic peoples have both maintained an unbroken narrative of anomalous pregnancies. The stories surrounding the births of Isaac, John the Baptist, and Jesus all fit within that pattern – a narrative phenomenon, far older and wider than the Judaeo-Christian tradition. Abrahamic, Greek, Egyptian, Norse, Celtic, Indian, and Chinese cultures – to name but a few – have all carried stories of *"Star Children"*.

Lao Tzu the philosopher and founder of Taoism, Emperor Taizu of Liao, and the Yellow Emperor would be examples of *"Star Children"* in the folklore of China. In each of those three cases the mothers were said to have fallen pregnant following strange encounters with objects of light illuminating them from the stars. These stories describe in words exactly what Carlo Crivelli depicted in his painting from 1648, as a UFO with its laser beam of light focussed on to the Virgin Mary's curious headgear. Mitra in Indian mythology – Mithras in the Graeco-Roman tradition – carries a similar story. Still in India, both Krishna and Vipassi, the twenty-second Buddha in the Buddhavamsa, repeat the same pattern. All these accounts speak of children whose conceptions followed close encounters experienced by their mothers – experiences of contact with extraterrestrial phenomena.

Of course, it's easy to imagine that such stories might be invented as a way of adding a measure of mystique to a religious authority, a king or a prince in ancient times. However, this rationalisation falls to the ground when it comes to more low-profile claims. For instance, when Akua from Anloga confided with her family about her Mami Wata experience, or when Jane Pooley on Australian National Television gave her account of ET insemination, neither would have expected any kudos in return for their story. Quite the opposite. What do we do then with such a widespread and longstanding testimony? What may have added mystique to a philosopher's CV three millennia ago is more likely to get a mother flagged for psycho-effective medication in the twenty-first century. Hardly a motivation to come forward with such a claim.

To the modern ear the language of "Star Children", what some call "Indigo Children", may sound like the language of fable and magic. It isn't. Think *in vitro fertilization, artificial insemination, genetic engineering, genetic modification, micro-surgery in utero, or hybridization.* Today we have language for technical phenomena which our ancestors could only view as supernatural. The question is, when we hear the report of analogous phenomena in our ancient texts, do we believe it? In a later chapter I will introduce you to John from California. John is a three-and-a-half-year-old boy and the vocabulary with which he brings his story is the language of a three-and-a-half-year-old. When you hear his story, you will have to ask, as I did, if you believe what he has to say. And if you do believe him, what then is the implication for the historic figures I have just named?

The Pig and Whistle – Canberra, Australia
September 2020

"If you say the word 'aliens' it's like a trigger for all kinds of cartoonish images. But in Sodom you've got two Sky People wandering around

the city centre, who could clearly pass for human beings. And not only that, they are so attractive that they get mobbed by the locals wanted to have sex with them! I guess if you can imagine the kind of hysterical mob that might form around a global pop star or a movie idol, it's that kind of reaction we're looking at. What do you make of that?"

"Well, Paul, I guess that explains how these Sky People managed to talk our ancestors into hybridising with them!"

Brad is humouring me. He and I are enjoying an evening of beer and aliens at a favourite watering hole in the Australian Capital Territory. Brad and I have been friends for thirty years. He is someone who brings a sober logic and an unfailing dry humour to any matter being discussed. He's also a good drinking buddy. This evening I am doing my best to massage the possibility of ET neighbours into Brad's general worldview, but I don't think I've convinced him.

"Paul, that story from Genesis is just so bizarre, I really don't think you can take anything from it. I mean, in the story of Sodom you've got a dad offering this rabid crowd the use of his own daughters instead of his visitors. I mean really; what are we looking at? What kind of story is that?!"

On that point I have to agree. The morality of the Sodom story really is bizarre, and it would be a more ingenious preacher than me who could extract a positive message from that text.

"I agree with you, Brad, but doesn't that absurdity tell you that what we're reading is not a moral tale? It doesn't aggrandize anyone's royal family or ancestry. I mean, no one comes out of that story looking good! Not even the Sky People. So, if it's not that kind of story, then what is it? What if that text is carrying a memory? Because so many of the stories of beginnings are like that – whether you're looking in the Bible, in the Sumerian stories or anywhere else. When I read them, it's like seeing the flashbacks of a patient with amnesia. You've got all these flashes, and glimpses and you can't quite see how they join up – but they are real memories. And if the only standout thing about Sky People, compared to the rest of us, is that they're a bit better looking;

if that's the mark of an ET presence, how easy would it be to miss?"

This proposition, at least, has Brad thoughtful for a moment. Another glass of Hefeweizen brings him some further clarity.

"Paul, do you think it's possible that these Star Child stories might just be a way of mythologizing the special charisma of figures like Buddha, or John the Baptist or Lao Tzu, as a way of dramatizing the feeling that people felt around them? When we experience that kind of charisma, or when we see the power of the movements they begin, is it that we feel we have to find a story, or make one up, to explain it? So they have to be born under a shooting star or at the conjunction of this or that constellation. Or they have to be half-human, half-God."

"I mean didn't the Romans use exactly that kind of story to get ordinary Roman citizens to worship the emperors? They probably heard stories of the gods of other cultures and thought, 'That's a good one! Let's use that story. We can adapt it to pump up the story of our empire.' Same thing in China. The Chinese emperors were divine, weren't they? Isn't it more probable that's where this idea of 'Star Children' or 'Indigo Children' came from? Or, conversely, it could even be the opposite: a story to make ordinary, totally non-famous people feel a bit more special?"

I have to concede there is a good deal of logic to that.

"That's fair, Brad, but what about the Jane Pooleys of this world who have nothing to gain from saying, 'Something funny happened to me just before I got pregnant, or while I was pregnant?' That doesn't fit your pattern does it? If we were really willing to stop and listen without judgement, and simply add up all the women all around the world who have stories like this to tell, then I don't think we have any easy explanation. I mean, just off the top of my head I can think of people I know on three different continents whose families carry stories like that. But they're not doing anything with that story. They don't brag about it. They're not making money from it. In fact, it's something they hardly speak about at all because they know people would think they're crazy. That little boy I mentioned, for instance, John from California, that's not even his real name. He's not looking

for publicity. He doesn't gain anything from his story."

After a moment's reflection, Brad takes a philosophical turn. *"At the end of the day,"* he says, *"the world is full of all sorts. We can't always understand each other's stories, or why we tell them – and that is a fact. I mean I have met many beautiful people who I've found enthralling. I can't always explain why. And it might not mean anything, but I might obsess about who this or that person was, what was their story, and what might have happened if I'd summoned the courage to speak to them. Doesn't mean they're an ET though, does it?"*

On the light rail home from the *Pig and Whistle*, I am feeling pensive. Brad is good at keeping my feet on the ground and I have no problem with accepting his points, which are all very reasonable. In fact, I am actually very grateful that our differences in worldview haven't dimmed our friendship in any way. In an era of angry and polarised public conversation, I find that having a friend you can disagree with is something very special. Yet the main difference between me and Brad is that ever since publishing *Escaping from Eden* I have been hearing testimonies every week from people who not only have nothing to gain from their stories, but who in many instances have been burdened and isolated by anomalous encounters they believe to be real, but which they know friends and family will not be able to process. It is a profoundly painful thing. Even more painful is to have experienced something you don't even understand yourself. My own worldview has shifted on the one hand because my studies in mythology have forced me to take stock of some unfamiliar things, and on the other hand because I have to find a way to process the great volume of first-hand experiences to which I have now been exposed.

As the light rail buzzes me through the streets of Canberra's north-eastern suburbs, I think again about the violent population of Sodom and how they perceived their visitors. They saw two powerfully attractive people walking through their city, and although magnetized by them they had totally failed to notice that these beautiful people weren't human. Something about that story takes me back to an evening in my youth. It's Springtime and I am twenty years old.

Chichester, England
March 1985

The night is dark but unseasonably warm. I am in Chichester, visiting my friends Rich and Trace. On my way to their college digs, I need to pick up some wine and some super healthy snacks from Holland & Barrett, the local health food store. For a specialist store the place is surprisingly busy. There are about ten people spaced out through the simple U of the floor space. But there's something very strange about two of them. Like a magnet my attention is drawn instantly to two very tall individuals. In fact, they are a couple, and the woman is pushing a stroller. They are light-skinned, with blonde-hair, and they are both incredibly attractive. As a twenty-year-old male I am not usually taken with the attractiveness of married couples with toddlers, but these three are simply the most perfectly beautiful people I have even seen, with an athletic, perfectly chiselled, Scandinavian look about them. Not only are they physically stunning but all three are radiating the most powerful aura of calm and peace and ease and... I can't describe it, but the feeling they are projecting is elating.

Now I notice something very odd about the way they are shopping. The woman has gone on ahead of the man and is taking things off the shelf and putting them into a basket. The man on the other side of the store is doing the same. They never speak and yet somehow they appear to be coordinating their shop. And there is no fumbling along the shelves for the items they're looking for. They both reach straight for each item, apparently without looking or thinking, and finally arrive at the checkout together. No one shops like that! It is almost as if they were either communicating without words, or as if they were pretending to shop, all the while throwing off this incredible... energy. I watch them, completely fixated, and I don't understand why. Why are they affecting me like this? What have they got? What's their secret?

For a very fleeting moment it occurs to me that these unusual and beautiful people might not be human. What then? Could they be angels? This thought lasts no more than a split second as the presence of the toddler immediately quashes that theory. I've certainly never heard of a toddler angel. I guess they must just very tall, beautiful, shiny, serene, charismatic people. Yet I cannot shake the feeling that there is something very different about them. The man then pays for their purchases without a word and they leave the store. I am next in line so I make my purchase as fast as I can, weighing up if I can catch up with them and somehow just ask them who they are or what they're on that gives them... whatever it is they've got. In the end, though, I decide that really would be too weird. And as it happens such a conversation isn't an option as the three beautiful strangers have vanished into the night.

It was such a strange experience. In the telling, it sounds like nothing and until telling you I have hardly ever mentioned it precisely because it doesn't sound like anything. I mean, what have I told you? I once saw some tall, blonde people who gave me a funny feeling. Why? I don't know. It's hardly a scoop! But today as I join the dots between stories – the Hebrew stories of Abraham's visitation and the visitors to Sodom, the stories of "those unlike us", or "the other crowd", or personal stories of Tall Whites or Nordics from people like my friend Patricia in Massachusetts – I now realise there are probably a few categories of being besides human or angel in the great diversity of a populated universe.

It's not an option I would have considered prior to the roller-coaster ride of researching and writing Escaping from Eden. Even now I hesitate to mention it because a perfectly reasonable reader might say, *"Paul, if you are willing to consider that ETs might be indistinguishable from tall, blonde Scandinavians, then clearly you're in a world where anything is possible!"* And I totally understand if that's your reaction. But I have two reasons for sharing this story that sounds like nothing. The first is that so many courageous men and women have taken risks with me, sharing things they fear might sound stupid, uneducated, credulous or crazy. If so many are willing to be vulnerable with me, then I am willing be vulnerable with you. I saw something I didn't understand, and still don't to this day. The second reason I am willing to tell you a story that sounds like nothing is that I want to encourage you personally to reflect on your own *"still can't my head around it"* experiences. What if you were to take a risk and share your story in the context of your family or friendship circle? How many other anomalous experiences will you hear in return? The more we permit each other to share our respective stories and the more willing we are to listen with an open ear, the more obvious it becomes that there is no shortage of real world phenomena beyond the familiar boundaries of our conventional explanations. If, having shared and listened, you find that your own worldview cannot accommodate the anomalies that you or a family member or a close friend have experienced, just pause and take a moment to ask yourself what precise boundary in your worldview is being challenged. What is the possibility you have not considered?

In the past, my worldview simply had no language for the three anomalies I experienced when I was twenty years old. I believed we were alone in the universe, so I had no label for the things I saw in my flat in Bath that night. I had never heard of *"missing time"*, and could only puzzle at those two glitches in my memory that year. On top of that, the only *"Nordics"* I had ever heard of were actual Scandinavians. So, Holland & Barrett simply had to remain a story about nothing. The Danish philosopher Soren Kierkegaard once said, *"Life is lived forwards and understood backwards."* I think he's right. Indeed, I now think that if we reach the end of our lives with the same worldview we learned in school, we surely have not been paying attention. Without such a journey, and the open-mindedness which should grow from experience, I would have had absolutely no idea what to do with the personal testimony of John from California, a three-and-a-half-year-old boy with a story that will send a chill down your spine. But that is for another chapter.

Chapter Twelve

Finding the Words

Cape Canaveral – February 1994

The noise of the engines roars across the waters to where we are standing, and hits us with a visceral force. I can feel the vibrations of it in my belly like I felt the power of the Horseshoe Falls, standing by the drop at Niagara only a few years before. It is an unnerving sensation of physical contact with the frequencies of immense energy. My body is resonating in the same terrifying way as the Space Shuttle *Discovery* rises from the launch pad on its pillar of smoke and fire and rockets into the cold, morning sky.

It is eight years since the world wept at the tragedy of the *Challenger*, and though some confidence has returned to the enterprise of shuttling payloads and personnel between Earth and orbit, not one of us in the crowd can be in any doubt concerning the essential danger of what we are looking at. *Discovery* is the shuttle which, in 1988, followed after the ill-fated *Challenger* to relaunch the American space program. As we watch her this morning, we are all mindful of the incredible courage of her crew – and of any crew willing to sit immediately adjacent to what is essentially a phenomenal explosion. How must that controlled explosion sound from the inside? It is both an inspiring and a fearful thing to see. For the sake of safety, the officials closest to the launch site are situated a full three miles away, sheltered by a building of reinforced concrete. Watch from any closer, or without the reinforced concrete in front of you, and it may be the last thing you see.

Getting off the planet is, apparently, the hardest part of the equation and the fundamental technology we use to do that is now more than seventy years old. It baffles me that we are still employing this brutal and potentially deadly technology, now so long in the tooth. And yet, as my mythological journeys have shown me, there is language in our ancient texts, suggesting that technology, looking and sounding uncannily similar, is in fact far longer in the tooth than we have ever imagined. It's there in the heritage of ancient carvings in Egypt, Central and South America, and it can be found in the pages of the Hindu Vedas. The presence in the Bible of such heavy-duty technology may be unfamiliar to many. Yet as I move from Genesis into Exodus, I find it is only a translation away from being blindingly obvious!

Language and linguistics were my first love. School and University taught me English, French, Latin, Italian, Portuguese, and New Testament Greek, and nurtured my enthusiasm for translation and interpretation. In time, this linguistic background gave me a useful foundation for touring the fascinating texts comprising the Hebrew Scriptures and the Greek New Testament. From these sources I preached quite happily for thirty-three years, although not without noticing a steady stream of anomalies which appear to belong to a world far removed from the beliefs and interests of orthodox Judaism and Christianity. Having now undertaken the exercise of rereading the Hebrew Scriptures with the word *elohim* translated in the plural, those anomalies have become easier to identify for what they are. For instance:

- An *elohim* story of genocide is a story of genocide – not a divine judgement.
- The arrival of a *ruach* over the flooded planet is a close encounter – not a divine spirit.
- The conflicts of *"The Fall"* in Genesis 3 are not about apples, snakes, sin and punishment and God's failure to anticipate the obvious. They are about a battle among genetic engineers over how intelligent human beings should be.

- The destruction of Babel is not about divine punishment for infringing ancient building codes. It is the obliteration of a technological, spacefaring civilization by an extraterrestrial force.
- In the book of Job two *elohim* toy with the life of a human being to test how powerful a hold the senior *elohim* has. It's not the story of a loving God. It is more like watching naughty boys burning ants with magnifying glasses. The total lack of empathy is exactly the same.
- The battles of the Old Testament are not stories of worship and apostasy. They are about the conflicts of extraterrestrial overlords, governing over separate human colonies and sparring with one another for hegemony and territory.
- When in the book of I Kings YHWH uses the word *elohim* to describe himself and spits on the name of a neighbouring *elohim* we are allowed a glimpse into the world of the ancient Hebrews before the story of monotheism came and airbrushed that forgotten world away.

For another example let me tell you something about the word *"glory"*. It renders a Hebrew word *"kbud"*. In the worlds of synagogue or church we are accustomed to hearing the word used as if it means the shining splendour of God's presence. However, as anyone who has preached on the book of Ezekiel knows, there's a whole other layer to that story! In the book of Ezekiel, the writer gives us a date, sometime in the C6th BCE. He then tells us what the *kbud* looked like when it landed by the Kebar River.

When he first notices it, the *kbud* is in the sky, surrounded by brilliant light, flashing lights and a cloud of smoke. Ezekiel has never seen anything like it and is struggling to describe it in terms of Earthly phenomena. What he can describe directly are jets of fire, wheels, metal, and glass. He also tells us about the noise it produced. When the *kbud* moved it made a sound like a mighty waterfall.

The word Ezekiel reaches for to introduce the pilot of the *kbud* is the word *"echie"*. It means *"life-form"* or *"animal"*. When the life-form pulls him into the craft (*"ruach"* in Hebrew) Ezekiel gets a closer look at the unusual pilot. He now says the pilot is *"like a human being."* In turn the pilot addresses Ezekiel as *"Human"*, and takes him on a flight around Iraq. The human-like life-form is evidently very interested in the religion and politics of Ezekiel's world and discusses both as they fly from place to place until finally depositing Ezekiel in Tel Abib. Once on solid ground, Ezekiel is so overwhelmed by the experience that he is unable to speak for a whole week after the encounter.

Today we have language for what Ezekiel reports. We would call his encounter a close encounter. The life-form – the human-like pilot – we would call an ET, and the journey an abduction. Ezekiel's life is never the same again.

This passage in the book of Ezekiel is the clearest explanation within the Hebrew stories of what a *kbud* actually is, and Ezekiel describes it for his reader in graphic detail. Etymologically a *kbud* is a *"heavy thing"*. Its behaviour in the story reveals that this particular heavy thing can carry at least two people and fly. It is a *ruach*, a craft, and Ezekiel uses that word too to describe the *kbud*. His very detailed physical description of the craft further confirms this very material usage of the word *ruach*, by telling us that the *ruach* is on wheels and that it transports people. In fact, the detail Ezekiel provides regarding the wheels of the *ruach* is so precise that Josef Blumrich, as NASA's Chief of Advanced Structural Development, was able to obtain a patent of the design for NASA. The patent, US3789947A, was issued on February 5th, 1974. It is called the *Omnidirectional Wheel* and is used by NASA to this day.

Sanskrit texts from the ancient Indian world of the Vedas have no embarrassment in describing the Vimanas in their ancient stories in technological terms as flying craft, with capability for both aerial and space flight. A text from between 1000 CE and 1055 CE, called the *Samarangana Sutradhara*, written by King Bhoja of Dhar, describes the capabilities and propulsion system of the Vimana. The Mahabharata speaks of vimanas which could fly, despatch sound-seeking missiles and destroy targets with what we would call lasers. Again, the ancient writers reach for metaphor to describe unfamiliar and powerful technology in Earthly terms. Some Vimanas are described in terms befitting Earthly palaces. Yet there is no attempt to make mechanical devices or physical technologies sound like something ethereal or spiritual. Read in that light, the appearance of YHWH's *kbud* on Mount Sinai takes on a different significance. The book of Exodus chapter 33 tells us this:

> *Mount Sinai was covered with smoke, because YHWH descended on it in fire. The smoke billowed up from it like smoke from a furnace, the whole mountain shook violently and the sound of the trumpet grew louder and louder. Then Moses spoke and the voice of the Elohim answered him.*

What then follows is a code of laws with which the Hebrew people are now to be governed by the Elohim in question.

At the end of a long conversation Moses asks if he can take a look at YHWH's *kbud*. In reply YHWH tells Moses that seeing the heavy thing at close range would be fatal. Instead, YHWH says, he will cause *"the goods"* (*"tub"* in the Hebrew) to pass by in front of Moses. YHWH also uses *"the goods"* to denote his *"heavy thing"*. Moses will then be able to watch the *kbud* as it moves away. However, to avoid being killed, Moses will need to be sheltered by a cleft in the rock and protected by YHWH's *"hand"* or *"pan"*.

This ever so intriguing episode has been rendered in conventional translations as Moses being allowed to see God's *"goodness... from behind"* but *not "face to face".*

The conventional translation raises far more questions than it answers. How, exactly, can you see *"goodness" "from behind"*? Furthermore, since Moses has been enjoying a face to face conversation with YHWH for some days, how can he now be told that a face to face encounter would prove fatal? Clearly something else is intended.

The explanation that YHWH gives to Moses is that if he sees the *kbud* it cannot be *"paneh".* The first uses of this Hebrew word *paneh* show it meaning *"on the surface"* or *"out in the open".* By that interpretation, what we are being told is that to avoid being killed, Moses must not be *unsheltered* or *out in the open* when the *heavy thing* moves. Given the dramatic physical impact of the *kbud* on the mountain, this warning makes perfect sense and the sequence of the story bears that translation out. It is precisely why Moses needs to shelter in the cleft of the rock when the *heavy thing* moves.

What Exodus has recorded – and what the translators have intentionally or unintentionally obscured – is a set of safety instructions relating to the launch of a powerful craft in the C13th BCE. Perhaps through having no technological grid by which to interpret the story, generations of translators have taken turns to morph it into the story of a deity who is so holy that no human being can see the deity face to face and live – except for Moses, who apparently can talk to him face to face – except for when he can't. And when he can, he must only look at God's *"goodness"* – but only *"from behind".* Does that make any sense? Any reader encountering this kind translation will know straight away that something is off. My view is that what cannot be seen face to face is not GOD but a big, heavy craft which belches fire and smoke with explosive force every time it lands and launches. Yet, somehow, we have been persuaded by translators and devotional teachers of the Bible not to notice when technology is being described.

We now travel four centuries into the future, arriving on the West Bank sometime in the C9th BCE. Here we will be treated to another close encounter with YHWH – this time as experienced by the prophet Elijah and his eyewitness, Elisha. According to I Kings 19, when YHWH arrives he is preceded by a powerful wind which tears into the mountains and shatters rocks in its path. The mountainside shakes and fire shoots out from whatever YHWH is arriving in. A few chapters later Elijah is taken into the craft in which YHWH has noisily arrived. The book of II Kings 2 describes Elijah being taken up into the sky, carried in a *"chariot of fire"*, through what appears to his friend like a whirlwind. Dare I say we have language for that in the C21st. Think *UFO, wormhole, abduction*.

Perhaps the most famous incident in the Hebrew scriptures concerns Enoch, the great-grandfather of Noah. An enigmatic verse in Genesis 5 tells us, *"Enoch walked with the Powerful Ones. Then he was no more because the Powerful Ones took him away."*

Can you say the word *"abduction"*?!

My experience as a preacher is that if you raid Biblical texts only for devotional lessons you will tend to skip past odd moments like these, or find some kind of *"moral of the story"* to apply to your audience. However, it is only a matter of time before the mismatch of religious language for extraterrestrial and technological phenomena becomes impossible to ignore. At least, that's what I found when I chose to pick up the gauntlet thrown down by the Vatican's senior astronomer Reverend Doctor Guy Consolmagno, when he issued the challenge to reread the Bible with a mind open to extraterrestrial possibilities. Without the language of ETs, UFOs, close encounters, abduction and hybridization, the Bible's stories are opaque and baffling, and when confused with God-stories they are frightening and deeply disturbing. With the admission of the extraterrestrial aspect, the phenomena may be hard to believe, but not hard to understand.

For me these points now seem glaringly obvious and it makes me wonder. If an eminent translator of Hebrew such as Mauro Biglino can identify ancient technology in Biblical vocabulary, there must surely be other Hebrew scholars, even within the walls of denominational headquarters, who know what he knows. Do they know in secret and agree in secret, while still dispensing the old, spiritualised translations? The answer to my question comes from a surprising quarter.

Canberra, Australia / Yorkshire, UK / Interlaken, Switzerland
September 2020

Our special guest today is nothing short of a living legend. He has been awarded honours in Peru, and Brazil and received an honorary doctorate for his life's work from the University of Bolivia. He has been continually in print for fifty-six years since his first publication in Der Nord Westen *– a German-speaking newspaper in Canada in 1964. Since then he has written forty-one books, which have been translated into thirty-two languages and has sold more than 72 million copies worldwide. He is the legendary Erich von Daniken.*

We are live on *The 5th Kind TV* hoping that the technology connecting Canberra, Australia with the UK and Interlaken in Switzerland will hold steady and record our interview with the channel's most famous guest to date. I am also a little hesitant, not knowing how Erich might have received George Noory's description of my book on air as: *"This generation's* Chariots of the Gods. *"* In 2020 at the age of eighty-five, Erich von Daniken is still a force of nature, and there is no doubt he is this generation's Erich von Daniken!

I need not have feared. Erich is warm and encouraging and as enthusiastic as ever as we delve into the subject at hand. I take our conversation to the beginnings of Erich's journey on to extraterrestrial territory. As a young, devout Catholic, his education was at a Jesuit school, where the study of languages required him to translate Biblical texts from one language to another. It was in these exercises that certain anomalous words began impressing themselves on Erich's young mind.

"There are about ten key words in the Bible," he says, *"which, if you retranslate them with a more literal meaning, they totally change the context and reveal the extraterrestrial story."*

My own studies in Genesis and the Hebrew Scriptures have brought me to the exact same conclusion. Why, then, do the old, spiritualised translations still persist? Is it purely a function of a closed worldview guiding translators' choices? Or, does Erich von Daniken believe the more etymological approach is avoided in order to keep a lid on forbidden knowledge? Are Mauro Biglino's former colleagues in Rome deliberately avoiding technological translations? In other words, does the Vatican know entirely more than it is telling?

"The translators of the past were brilliant human beings," he says. *"Highly educated. People of Integrity. It was not the Zeitgeist of today. They translated the texts according to the thought-forms of their time. But now it is time to make some changes in how we translate the old texts."*

For Erich von Daniken the gradual drip, drip of disclosure from the Catholic Church is really just a matter of revealing what the public is ready to accept as the general understanding of society progresses. Previous generations had no points of comparison to enable them to recognize technological moments in the Bible. By contrast, in the C21st, we now know about space-faring technology, the movement of planets, wormholes, asteroid impacts, artificial insemination, genetic modification, hybridization, telecommunications, universal translators, blue-tooths, and spacesuits. All these things are now a familiar part of our world. Consequently, we have knowledge today which allows us to understand phenomena, recorded by our distant ancestors, which would have been utterly incomprehensible to intervening generations of translators. Without any disrespect to previous generations of translators, we should not be afraid to say so.

However, for any person of faith, Moses' encounter with a YHWH who uses smoky, fiery technology is one of a number of Biblical anomalies that point to a far-reaching question. If a moment so central to the story of the Bible turns out to be something other than a GOD-story; if the YHWH of Moses was no more than Israel's particular Powerful One, then where does the real GOD actually turn up in the pages of the Bible – if at all? It's a tricky question to answer. The C6th BCE redactors of the Hebrew story have deliberately made it so. The whole purpose of their final edit was to make the diversity of stories gathered together in the Hebrew Canon appear like a seamless story of GOD – the Source of all things – going right back to the turbulent stories of beginnings.

However, the erstwhile Vatican translator Mauro Biglino argues that GOD makes not even the briefest appearance in the pages of the Old Testament. For him, the whole story is one of human subjection to ancient Powerful Ones.

Let me tell you where my thinking has reached. Wherever I see YHWH appearing as a character in the story, playing an active role, I believe we are looking at a Powerful One. He is often punitive and brutal, sometimes unpredictable, and can be implacable and unforgiving in the extreme. In those texts YHWH appears to govern the tribes of Israel in the same way the other *elohim* govern their humans. Yet within the Hebrew Canon there are glimpses of something higher. For instance, Amos was a Jewish preacher of the C8th BCE. His vision of GOD (who the text calls YHWH) is of a loving intelligence, the creative source of the universe, with a vision of love and justice extending to all humanity – including Israel's enemies. To me, that sounds less like the brutal *elohim* overlords and a bit more like GOD – the harmonious Source of all things – whom Jesus addressed as *"Father"*.

In the Gospel of John, we can find that same idea of GOD as Cosmic Source reaffirmed. The Gospel begins: *"In the beginning was the word. And the word was with GOD and the word was GOD. The same was with GOD in the beginning. Through the same all things came into being... In the same was life and that life was [and is] the light of human beings." (John 1:1-4)*

The Apostle Paul in the mid C1st CE spoke in similar terms, when he said, *"I am speaking to you about GOD – the source of the Cosmos and everything in it." (Acts 17:23,24)*

Where then does Jesus see this Cosmic GOD in the Hebrew story? If we go looking for Jesus to affirm the claims about GOD made by the monotheized story of the Hebrew Canon the search is not as revealing as one might expect. In the Gospel according to Matthew and Mark, Jesus often repeated a formula that said, *"You have heard it said... but I say..."* or *"Moses permitted you this... but I say this..."* This form of words puts a clear distance between Jesus' vision and that of the old, old stories.

In one place Jesus affirms a command from the law of Exodus as being a word from GOD. It is the command to look after your parents in their old age. It is one of a small number of sayings in which Jesus refers to the God of his hearers in such a way as to blow up their whole theological framework. Put simply, he is citing his audience's own theology against them. By my reckoning, anyone looking for Jesus to summarily affirm *"the God of the Old Testament"* or *"the Hebrew vision of God"* is going to come away from the Gospels with thinner pickings than they might expect.

This is the reframing that my journey into the depths of Genesis forced upon me as I sat in my shipping crate cabin at the leafy end of my driveway in Victoria, surrounded by the texts of the Bible and the parallel accounts of so many cultures through the ages. As the picture I have outlined in these pages took shape for me, I began to realise that if I was to stand up for my findings I was really going to have to take my life in my hands, surrender any thought of popular acceptance in more religious circles, and be willing to live with the impact on my reputation that such conclusions would inevitably occasion. And to be very honest with you, even now I find there is a real embarrassment threshold I have to cross every single time I allow the language of ETs, close encounters, abductions *etc.* into my vocabulary. The impulse to self-edit for the sake of social acceptance is deep wiring. So I cannot help but be conscious that any hearer is likely to judge my sanity the moment I include such words in a sentence. Yet embracing such controversial language has enabled me to understand more clearly the cultural memory curated by our ancient and sacred texts. Having taken up his challenge, I find that Reverend Doctor Consolmagno is quite correct. In the Bible the ET presence really is from start to finish.

But now I am really scratching my head. My level of education is the average for a working pastor – a Bachelor's degree with Honours. The scholars whose work produces our Bibles and Bible dictionaries have multiple Masters and Doctorates. Where are they in the picture? If I can see the case for extraterrestrial and technological aspects in the Bible; if a highly skilled and qualified translator like Mauro Biglino can see it; if an eminent Jesuit scholar, a senior Vatican theologian like Guy Consolmagno can see it; if a teenage schoolboy in Switzerland by the name of Erich von Daniken could see it; there must surely be highly qualified academics, who behind closed doors can also see these possibilities. Where are they in the mix? Are they all in the closet? While I have Erich von Daniken on the line, I take the opportunity to ask him if he has ever been approached by academics who actually support extraterrestrial and technological translations in private but who, for whatever reason, find it impossible to make their position public. He beams as he replies.

"Paul, many, many academics have come to me – especially those who have retired. Sitting together over a glass of wine they confess, 'Erich, I hold a similar view to you, but because of my position I could never make it public.' Some have given me material to publish and I would say, 'Thank you, Professor. This is great. I will certainly use this information in my next book and of course I will credit you.' And they say, 'No, Erich, don't. You must never mention my name! It would damage my reputation. I would lose my job.'"

I ask Erich if it frustrates him to have his theories affirmed by people in secret who then refuse to go public with their views. He says, *"No, because I have received so much information that way [to support] my case. When authorities tell me, 'I will give you this information but please don't mention my name,' I always respect this. Otherwise it would break the relationship. It has to be the outsiders who bring this kind of discussion into the academic community. Then it can be discussed – but the idea always has to come from the outside."*

It intrigues me that I haven't picked up from Erich any kind of a gripe towards the monolith of Roman Catholic orthodoxy. Far from it. In fact, he tells me, it was a Jesuit educator who first encouraged him while still a teenager at school to nurture his extraterrestrial questions with a visit to the Book of Enoch. He says,

"The Jesuit brothers who taught me, every single one of them had at least one doctoral degree and they were very, very open. We had no fights and later when Chariots of the Gods *came on the market, many of my lecturers had retired. But some of my lecturers were there at my [book launch]. After my speech we were sitting together, drinking a glass of wine, and they all congratulated me on a wonderful result!"*

More than half a century later the congratulations are of an even higher order. With great enthusiasm Erich holds up a letter for me to see – a beautiful parchment, embossed with the grand wax Seal of Pope Francis. It is a Papal gift, congratulating Erich von Daniken on his eighty-fifth birthday. A very nice touch. Yet, somehow, Pope Francis' sincere and generous gesture makes me wonder all the more keenly, *"What do they know that they're not saying?!"*

Chapter Thirteen

Beginning to Remember

Canberra, Australia – May 2020

"I respect what you're saying, Paul – and it's quite interesting as well – all three of the things you said from when you were twenty..."

It is a pleasure for me to talk with a researcher as diligent, highly intelligent, and grounded as Richard Dolan. He is a gentleman, very gracious and polite, but is certainly not one to go along with flights of speculation and non-evidenced arguments. We are talking about the recent disclosures from the Pentagon, US Department of Defense, the Navy *etc.* The official acknowledgement of metamaterials is in the news – for those paying attention. So, we have been drilling down into those topics. Now we're on to more personal territory.

"The appearance of the entities in your room; I understand that it was a very clear vision that you had, so that you were quite sure that you were not dreaming, and that is why you are sharing it now. This is a story that other people have told me, both as children and as adults."

I am eager to bounce off an expert my own flashbacks of – whatever they were – those mystifying moments which had been accompanying me on my journey to *Escaping from Eden*. Richard compares what I have shared with him to the story of another experiencer in his circle.

"In one case that I know very, very well, it was three entities that were in a doorway. This person at the time was about eleven years of age. She saw silhouetted three entities in the doorway of her bedroom. The hall light was on, so she had this silhouetted vision – and she was adamant she was awake. It was not a dream. She closed her eyes and wanted them to go away. She opened her eyes and they were still there. This happened several times. And then she lost her memory. She woke

the next morning with a triangular series of pinpricks on her wrist."

"Like you, she always assumed it was some kind of a demonic or spiritual thing. It was only many years later that she wondered, 'Wait a minute, was that an ET visitation?' And in your case with that experience of missing time around the same period, it's a natural thing to wonder."

I think about the pattern of that encounter. The girl saw strange beings for just a few moments and then she didn't know what happened next. Next thing she woke up with a scar. It reminds me of Dean and his landscaping crew who saw a craft, then lost two hours, with one waking up to find creepy markings on her arm.

This phenomenon of remembering only a few fragmentary moments and then not knowing what happened next; I wonder if it is a widely recurring story. Barbara Lamb says, *"Yes."* In fact, it was a massive correlation of patterns just like that which first got the attention of this grounded, mild-mannered and gracious psychotherapist. It sent her on a journey of discovery which in time was to transform her life and work.

San Diego, Southern California
August 2020

Barbara Lamb is a licenced psychotherapist whose work in relationship, family and child therapy goes back more than forty years. In 1991 a young lady came to see her as a client with a disturbing and embarrassing problem she needed to discuss. She had previously been living independently and was enjoying her life of study as a university undergraduate. Now she had moved back in with her parents and was so terrified of something that she could only sleep in her parents' bed. It was her parents who drew a line and took the initiative to seek out some therapy. By one of those strange synchronicities the universe offers us from time to time, the young lady's mother bumped into Barbara in a bookstore and recognized her as a regression therapist. This was the encounter that drew Barbara into the case.

Once in her counselling room, Barbara gave the young lady some exercises to help her get into a state of deep relaxation. As she spoke, memories which her mind had buried in its subconscious began to re-emerge. The memories involved night-time visits from non-human entities who would take her somewhere, examine her, and then return her to her room. Prior to her first session with Barbara, the young lady had only been able to recall the first few seconds of each encounter. The rest had been a blank. What she had remembered was enough to leave her terrified.

Thankfully there is a happy ending to the story. After six sessions the young lady returned to tell Barbara that she felt relaxed and happy, and was reconciled to the strange encounters, which she now accepted had left her unharmed. Her demeanour was so changed and her recovery of confidence so complete that she and her boyfriend had agreed upon a rural home out in the wilderness, without any anxiety over the retrieved memories.

"That was the beginning of my experiences of working with people recovering memories of extraterrestrial encounters."

The method Barbara employed to help her young client – and all those who followed with similar memories – was very close to that of her friend and colleague Professor John Mack, head of Harvard's Department of Clinical Psychology. John used a discipline of controlled, conscious breathing to enable his clients achieve a state of deep relaxation. He would then invite them to go back in their minds to the first moments they could recall of the encounter in question, and ask them what they could see.

During each client's report of what they were seeing, hearing and feeling, he would invite them to look to the left, or up, down or to the right and tell him what they could see. It was in their answers to these supplementary questions that Professor Mack began to notice an inexplicable correlation of incidental details. Inexplicable, that is, unless they had all been in similar places and seen similar things. It was these correlations that drew the Professor further into the rabbit hole to take a closer look. Barbara was one of the many academics and clinical practitioners who wrote letters of support to the board at Harvard when they started putting pressure on him for his conclusions.

"John and I were really good friends." Barbara beams at the memory of him. *"We really understood each other's work and had lots of discussions about that. In many ways my work was very similar to his. I was already doing this kind of work out on the West Coast and he was doing his on the East Coast. After 1991 people started coming with the complaint that they had been visited by very unusual beings in the night and were taken away for a while. They usually were very upset and even traumatised by this, and were at a point in life where they would like to know more details about that – and especially to know if their experience was really real. For many who are regressed there is a feeling of relief that they now know – after many decades in some cases."*

"They have wondered about those peculiar experiences that they have remembered just a very few moments of. And they have carried that mystery, that unease with them in all the time since. So when they finally come and do a regression and get the validation that they lived through the experiences there's a certain sense of relief that they feel. Whatever was buried in the subconscious part of their mind is now known to them consciously, and that means they can work with it, adjust to it."

I ask Barbara exactly how many of her clients have reported experiences like these in the thirty years since that first experiencer crossed her threshold.

"Oh, by now it's more than two thousand!"

Barbara Lamb has built up an almost unparalleled clinical record of cases of experiencers of abduction. Since publishing *Escaping from Eden* the mix of clients who find my website and come to me for personal coaching has shifted seismically and, like Barbara, I am privileged to have sat with many as they process their own anomalous experiences. Some of the experiences I hear related are decades old – such is the power of taboo to keep us silent. Some months I hear such reports every week. Some weeks it is every day.

This work has set me on a steep learning curve. In this potentially disorienting territory it is my knowledge of world mythologies that keeps me grounded and offers me insights to help bring context and understanding to our conversations. It is a deeply painful thing to carry the memory of an experience you don't understand. The moment you can understand what happened or why it may have happened the sense of relief is profound – even if the experience has left questions and uncertainties.

You might wonder where that gets us. If we can arrive at the point of accepting that this contact or that abduction really happened, what does that mean? What's it about? Why would extraterrestrial beings be interested in semi-covert contact with human beings? Why would they be interested in abducting human beings? Why would another similar species wish to hybridise with us? To a C21st mind it may seem a far-fetched proposition and yet as we have seen on our journey around the world, the hybridization narrative is probably the most widely recurring theme in our world mythologies of paleo-contact. When I began probing the book of Genesis for signs of an ET presence, the hybridization story from Genesis 6 was the one potentially extraterrestrial story I already knew about. The world knows the story from Greek, Norse, Indian, Chinese and Celtic folklore – and of course from the Welsh, Filipino, Caribbean and African cultures that have curated stories in the lineage of Mami Wata. One thing which has surprised me on my journey of discovery though is the calibre of researchers willing to nail their colours to the mast and say they believe it is happening right now.

Canberra, Australia
May 2020

As I compare notes with a grounded researcher of the calibre of Richard Dolan, I am surprised at how openly he is willing to make the following admission: *"The hybridization program, the program to extract genetic material from us, I think is probably very widespread. That could mean a LOT of people contributing to the program."* A few weeks later as we talk on camera Barbara Lamb unpacks what *"A LOT"* might mean.

"I think there are millions of people worldwide who have been having these encounter experiences. Most probably know only a little bit about it and wonder about it, and may be quite troubled about it. They can easily wonder about their own sanity, which of course adds greatly to their distress. It's an awful feeling to think that you're alone in something so peculiar and that you have no one to turn to."

It is a peculiar kind of narrative and no doubt about that. I ask Barbara what she thinks might be the motivations behind the kind of hybridization program evidenced both by ancient mythology and by the lore of today's experiencers and researchers.

"A lot of that has come up in the work I have done. All in all, there are about fifteen reasons that have been given to my clients by the beings they encountered. The reason that was given the most frequently is that those species of extraterrestrials who were creating hybrids were species who had come to a point where they realised that their species was having great difficulty in reproducing the next generation. Their civilization was in the process of slowly dying out and they wanted to rescue their species. They realised that they had to combine with some other sort of species in order to survive."

"Another species had gotten to know about human beings from afar and liked some of our characteristics. Though puzzled by them, they liked the fact of our emotions. Most of those species are very low in the

component of emotion or have no emotional reactions at all. They're just not constructed that way and they have noticed that humans enjoy a whole variety of emotions which make their lives interesting, colourful and creative. They respect that humans have a whole panoply of artistic expressions which make our lives stimulating and interesting. And they like the look of that."

"Also many of the beings have a very frail body form. They are very developed mentally but with weaker bodies. So one of the things about humans that they like and want to combine with is our more robust physicality."

All that Barbara says brings to my memory an intriguing verse from the Bible. It is the sentence in Genesis 6 that introduces the episode of alien abduction and hybridization by the *benei elohim*:

As the human population on Earth grew, producing daughters, ones like the Powerful Ones (benei Elohim) saw how beautiful the human females were. They chose whichever of the women they wanted and took them. The children they bore were the Nephilim – the powerful people – the ones of legend.
(Genesis 6:2,5b)

Evidently, there was something about human beings that these *"Ones like the Powerful Ones"* found especially beautiful. I wonder if the beauty that attracted them was more than skin deep. What Barbara says about love and creativity makes sense to me. And the Genesis story tells us that the result of this combination of species was a stronger, more robust kind of being. I don't find it hard to believe that there is something special about humanity, the unique fusion of animal strength, mammalian emotion and higher intelligence, that makes human consciousness something rare and special, and which might very well excite the interest of others in a populated universe. We only have to compare humans to every other species on this planet to appreciate that there is something awesome about the unique fusion of the human condition. There is both a finesse to animal consciousness and a uniqueness to human consciousness which when combined make for the incredible species we are, capable of such love and such beauty. Is it any wonder that other species would see something in us that might attract?

Those who contact me for coaching certainly do wonder about it. Young and old, professional people, scholars, technicians, researchers, medics, defence personnel, clergy; every experiencer wonders about what they have experienced as they try to wrap their heads around what has happened to them. They wonder if they have made the right sense of it; if the picture they have formed bears any relation to things others may have experienced or concluded. As Barbara says, it's an awful thing to think you're alone in something.

I count it a privilege to be trusted with people's stories. They don't tell me for money, fame or even attention. They reach out to me because, sometimes even many decades after their encounter or encounters, they still need to process whatever they have remembered of it and try to come to terms with what it means.

Often in our initial conversations, clients ask me about my own experiences. Have I ever experienced a close encounter? Do I have first-hand knowledge of an ET presence on or around planet Earth? In that context I am happy, for what it's worth, to share my own story even if it appears to be a hotchpotch of vague and fragmentary memories. Even now as I share these flashes and glimpses with you, like crumbs along on the pathway, I am putting the picture together for myself, and to tell you the truth I am not sure if even I want to go where these crumbs are leading.

Chapter Fourteen

When Predators become the Prey

Canberra, Australia – May 2020

The encounter in my flat in Bath that night has always bothered me. Something about it has always seemed ridiculous. My last memory of that encounter was of hiding under my covers, raspily rebuking the small grey entities as if I were in *The Exorcist*, and my heart pounding fit to burst out of my chest. In that state of agitation, how could I possibly have fallen asleep? It's a question I now put to Barbara Lamb.

"Most people remember just the first few moments of beings being there, and not understanding it," she says. *"You see the beings have a way of what we call 'switching us off!' And if you happen to be with somebody else – someone with you in your bed at night when these beings come, or somebody else in the car with you or wherever you are when this begins, that other person will be deeply switched off. There would be no way you could waken or get the attention of that other person. So it seems like the companion is switched off first – not harmed in any way but just made to be not conscious. The person having the experience will experience just the first few moments of their encounter before they too are switched off. That is why they don't remember what happened. So, Paul, what you describe is very typical of this kind of experience."*

In all the years that I have journeyed with that strange flashback from 1985 *"typical"* is not a word I would have associated with that memory. Indeed, as Barbara shares the report of clients describing their memories of close encounters and abductions, her language is surprisingly mild. Many of the experiencers whose stories she relates use the same kind of non-judgmental language towards their abductors. It's puzzling. The Australian nurse Jane Pooley even speaks in warm and affectionate terms of the beings who she describes as having periodically taken her for more than half a century. Her mild unjudging language is out of kilter with the horror we might expect to accompany the kind of experiences being reported. There is a psychological phenomenon called Stockholm Syndrome. It describes what happens when the victim of a kidnapping falls in love with their kidnapper. Some kidnappers deliberately exploit this phenomenon as a way of manipulating and managing their captives. Are Jane Pooley and many of the two thousand who have shared their stories with Barbara Lamb all victims of Stockholm Syndrome but on an even more dramatic scale?

Somewhere in New South Wales
June 2019

Hugo is barely more than a toddler when he is taken. One moment he is living a life of childlike innocence in the warmth of his family, the next a group of large and powerful entities have taken him. Frightened and disoriented, he finds himself airborne, crouching in the corner of some kind of craft which is carrying him at an unimaginable speed to an environment that is nothing like he has ever seen before. For the first few days Hugo is bewildered and frightened, trying desperately to find some way back to his family and the world he knows. But the experience is softened when one of the other entities appears to befriend him. Though he cannot understand the entity's language he can sense its intention. It is gentle with him and tries to comfort him.

What he does not know is that a generation before him, his mother went through the exact same process, as did her mother in the generation before. These beings have followed the bloodline of Hugo's family without that knowledge ever really being communicated from one generation to the next. As Hugo adjusts to the situation, he realises that none of the entities in this strange new place are actually doing him any harm. As time goes by, Hugo feels not only that they are looking after him but that they care for him. Sometimes the feeling of their love and care for him is overwhelming. At some telepathic level Hugo senses his captors' desire for him to be happy. Then one day there comes a moment when Hugo realises the advanced beings are giving him the choice either to remain with his captors or to leave. He chooses to stay. The captors have become his family.

Hugo is my cat. I am one of the advanced beings who abducted him from his maternal home in New South Wales. We meant him no harm when we purchased him from the breeder. We have always loved him and are very confident that he has come to love us. As humans we permit ourselves to believe that we have not been cruel by making him part of our family. This is our mentality towards pets. We understand a little of their emotionality. Indeed, that is an important part of why we love them. Yet at the same time we are willing for them to endure the pain of separation from their own species and their own family in order to become part of ours. Is it possible that another species might take a similar view towards human beings? Might we be like pets to some of our neighbours, like cattle to others, and like vermin to still others? Is it possible that within the spectrum of our cosmic neighbours there are entities who take us and exploit us like a pedigree dog or cat breeder – without anything they would perceive as a malicious intent.

I know that making this comparison risks offending a lot of people. I am deeply aware of the inconsolable grief of any family who have lost loved ones to they don't know who or what. To lose a child or partner or parent; these are the deepest hurts. To be overpowered by something we can't fight and don't understand; this is the greatest horror. There is no reframing in the world that can assuage the pain of that. Abduction is abuse. Nothing makes it acceptable. Today, in the C21st, we are just beginning to get a sense of scale of the issue of human trafficking and abuse around the world. It is so horrific that we hardly allow ourselves to look at it. My years as an Intentional Interim Minister and an Archdeacon involved me in the work of professional standards teams, and leadership bodies engaging with the Royal Commission into institutional responses to child sexual abuse in Australia. That work forced me to engage with some dark realities.

The human aspect of it is disturbing in the extreme. Yet if you take the time to join the dots between the successive aborted police investigations into ritual abuse and elite paedophile rings, the scandals surrounding presidents, royals, Hollywood moguls and all the rest of it; if you join the dots from there to the debased world of human and child sacrifice in the cultic practices of antiquity, stretching from Mesoamerica, to Mesopotamia to Asia; if you have any awareness of the Epstein-Maxwell case, or what the recent Government and Royal Commissions have uncovered in the UK and Australia; then you will have some sense of how interlocked, pervasive and ancient these patterns of abduction and abuse really are. To acknowledge that there may be an equally ancient, non-human aspect to the phenomenon is both disturbing and terrifying.

One of the reasons that perhaps most people will not even want to consider the possibility of a non-human aspect to the abduction phenomenon, is that our acknowledged experience as a species is that of being our planet's alpha predator. It is hardly a welcome idea that some other kind of entity may be our galaxy's alpha predator, or that some presence might regard us as pets or prey to be exploited or taken. Even the least violent of abduction narratives trespasses our most fundamental psychological boundaries, if it is our hope to maintain an illusion of safety.

The statistics for people who go missing around the world every year are truly horrifying. Whether the secrets behind these cases are ones of misadventure, serial killing, ritual abuse, sex trafficking, or something non-human, the frighteningly high numbers of missing people ought to be continually in the news cycle. Somehow as a culture we seem to lack the courage to keep the phenomenon squarely in the public eye. Perhaps the scale of it and the not knowing are simply too overwhelming. Sometimes, however, it is what those who return have to say that we don't want to do business with.

Northern California
September 2nd, 2011

John (not his real name) is camping with his parents and grandparents at Fowler's Campground, one of their favourite spots near the McCloud River. It's a favoured fly-fishing area on Mount Shasta. It is 6pm and time for dinner when John's parents realise their boy is missing. He was literally there one minute and gone the next. John's father immediately calls the police, and officers from the United States Forest Service. The local volunteer fire brigade has officers there within forty-five minutes and soon more than one hundred paid and unpaid search and rescue people are at the site, all looking for John. John's father is distraught and continues desperately searching until after five hours he collapses in an exhausted heap and has to be carried back to the camp. Before long the forest is pitch dark and the threats of the environment, which include bears, become all the greater. The prospects of survival for a three-year-old child, alone in a Californian forest overnight, are not good.

Shortly before 1am, the cold wet nose of a Dutch Shepherd dog nuzzles a small, frightened boy, sheltering under a bush. The dog's name is Tom and he is a canine officer with the Siskiyou County Sheriff's Office. The bush where Tom has found John is right next to the trail. And it's a trail that has been searched now for nearly seven hours. Yet there is John, unhurt, but frightened and dazed when they rouse him. After a check at the local hospital, John's mum and dad take him home in a state of exhaustion and unimaginable relief.

However, the story doesn't end there. A few weeks later John adds a disturbing postscript when playing one day with his grandma Kathy, who he calls *"Kappy"*. From out of nowhere, John looks up and tells his grandma, *"I don't like the other Grandma Kappy!"*

Naturally puzzled, Kathy asks John to explain what he means. John reminds her of the time when he got lost in the woods. He tells his grandma that he was taken by a woman he thought was her. He says, *"She had your same hair, your feet and even your face."*

John goes on to describe how the *"other Grandma Kappy"* took him into a cool, dark cavern. He recalls seeing some dusty old purses lying on the floor, and near the entrance some small guns and other weapons. Standing motionless around the cave walls were beings who looked to the boy like robots. Strangely there was a ladder inside the cave. The *"other Grandma Kappy"* then climbed the ladder to fetch something and as she did John saw that she was moving in a way that appeared mechanical. Then he noticed an odd light emanating from her head. Bit by bit it dawned on John that this figure was not his Grandma Kappy. Unsure what she is listening to Kathy probes a little further, *"What did she do with you, buddy?"*

"She made me lay down and looked at my tummy," John replies. *"Then she tried to get me to poop on a sticky paper, but I couldn't go. She told me that I am from outer space and they put me in my mom's tummy. Then she took me back to the river and said to wait under the bush until someone found me."*

It was certainly a bizarre story. Could John have fallen asleep in the cold, dark forest and simply dreamt it all? Could he have eaten toxic berries and hallucinated the whole thing? Three-year-olds certainly don't lack for imagination. However, Kathy has a personal reason to take John's testimony a degree more seriously, and that is that not more than one year before, something equally mystifying had happened to her in the same location on Mount Shasta.

Kathy was part of a small group of friends enjoying a weekend's camping. That is, until one morning Kathy awoke to find herself lying face down in the dirt, outside the tent, with absolutely no memory of how she got there. She felt weak and sick and her neck hurt. When Kathy's friends anxiously checked her over for any signs of harm, they found that the skin at the back of her neck was red and raised, with a puncture wound in the middle of the enflamed area. Another camper in their group, who had slept in a caravan on the site, found the same injury on his neck. It took them both some weeks before their health fully returned.

Something else from that trip had bothered Kathy and her friends at the time, and that was the surprising lack of wildlife at the spot where they camped. All the birds, squirrels or butterflies they would usually expect to see around them in this part of the forest were conspicuously absent, except that night Kathy's group did notice some red eyes gazing at them from the woodland. At the time Kathy's group assumed the eyes were those of forest deer, and that the puncture wounds were from spider bites. Kathy persuaded herself that the poison from the spider bite must have disoriented her and caused her to crawl in her sleep out of her swag and out of her tent, only to wake up in the morning. However, now that this has happened, in the same area, to her own grandson, and now that she has heard his story, Kathy is not so sure.

I first encountered John's case when Kathy relayed her experiences in a research forum twelve months after the incident. What did she make of John's story of having his tummy examined and being asked to defecate on to sticky paper? And what about the *"star child"* aspect of John's testimony? These are strange details to hear from a three-year-old. What did Kathy make of it all?

"I called my son and asked, 'What the H are you letting my grandson watch on TV?' And I told him what [John] said. He said that [John had] told them the same story a few days ago, but [they had] chalked it up to having the smartest, most amazing kid with the biggest imagination ever."

But, for Kathy, something about that explanation didn't sit right. John was now having nightmares, not about getting lost, but about extraterrestrials. She says,

"I know that kids have imaginations but... I think [John] is trying to tell us what happened in the terms he understands – or thinks we will understand... It was the pooping on a sticky paper that really makes me wonder. I've never watched a TV program that mentioned pooping on sticky paper [for John to have copied that]. There were other details too. Too much to list."

Kathy's experience with her son, her grandson and her own encounter illustrate what is the natural and rational response to anomalous experiences. We try and explain them in familiar and conventional terms – nightmares, hallucinations, childish fantasy, wildlife, spider bites, toxic reactions *etc.* In years past, if I had heard young John's strange account of a brief abduction, I would have done the same and would imagine that the boy must have experienced some kind of delusion, either a dream or a reaction to some kind of toxin. But after all I have heard from credible witnesses in the last few years, I have to wonder how many of us have unsettling stories which we have half-explained away yet have never fully settled in our memory.

What would happen if we were to sit down and allow each other to pool our half-explained encounters? I wonder. How would we deal with the picture that would emerge? As a culture, do we have the courage to name it and confront it?

Chapter Fifteen

Lifting the Lid

San Diego, California – 2020

The last time I was in San Diego I was fifteen years old and was here with my family to visit SeaWorld. Today I am on a mission. I may have been overzealous in my efforts not to miss my 2 o'clock appointment as I have arrived a full forty minutes ahead of schedule and now find myself nervously flicking through décor magazines as the time ticks slowly by.

I am feeling pensive. I can't stop thinking about that little boy, John, and his strange encounter on Mount Shasta. As I reflect on it, it strikes me that John's other-worldly encounter bears some uncanny similarities to the experience of my friend Juan Perez in Argentina all those years ago.

- Juan described the craft on his grandfather's farm as a *"hut"* and spoke of the beings inside as appearing robotic. John spoke of a cavern with robotic inhabitants.
- Like John's cavern, Juan's *"hut"* bizarrely featured a ladder – hardly the technology one would associate with an advanced species or a close encounter.
- John encountered an entity which projected the appearance of his grandmother. Juan encountered an entity who he interpreted to be his late grandfather.
- John's grandmother experienced a mysterious phenomenon of her own. Juan's mother also experienced a close encounter.
- John's grandma suffered a strange puncture wound and inflamed red skin. Juan's encounter left him with a strange red mark – like an inoculation mark on his upper arm.
- Both Juan and John were children at the time of their

respective encounters.

The details intrigue me. How much of this was objective? How reliable were their recollections? Certainly, children can have vivid imaginations. But they can also relay what they see with guileless innocence and directness. And the aftermath? Juan still suffered from PTSD forty years after the event, and in a conversation four years after his encounter on Mount Shasta, John's grandma reveals that he is still having nightmares. I don't think a child's made-up story would do that.

John's description of what the fake grandma had to say about his conception as a *"star child"* expresses in the language of a child the themes of the Mami Wata traditions of Africa and the Caribbean, the Diwatas and Dili Ingon Nato of the Philippines, and the fay traditions of Celtic Europe – a worldwide narrative of abduction and hybridization by non-human neighbours. If our worldview has no place for such neighbours, then any such claim is an assault – an *"insult to our intelligence"*. Speaking personally, I don't think John is insulting my intelligence.

Ultimately, worldview must adjust in the light of information. My own worldview has certainly had to adapt in the light of all that has surfaced through the last few years. I can only wonder what further shifts might result from this afternoon's session. I have never done anything like this before. What new revelations might await me? What new implications will I have to process?

Twenty minutes to go.

The first seismic shift in my worldview with regard to ETs and abduction reports like those of Juan and John came in 2009, when a worldview defining agency – the Vatican – laid down its public challenge for people to get ready to welcome the physical presence of extraterrestrial relatives. This was my first red pill.

Next came my linguistic discoveries in the book of Genesis and in our world mythologies which affirm ET interventions in our history. My second red pill.

Discovering Plato's world-changing claims, and the early Church Fathers' endorsement of Plato; that was my third red pill.

Take three red pills and there's no way you're going back into the matrix!

As I listened to the supporting voices of psychologists, government officials and other witnesses from the scientific community, each successive witness provided another nail in the coffin for my old paradigm – the paradigm that said, *"We are the pinnacle of evolution, the crown of creation, unique and alone in the universe. End of story!"*

Sharing the process with you has brought home to me how very fortunate I am to have been helped in my understanding by so many witnesses of another world. I think of Alan Stivelman and Juan Perez and the courage of their stunning film-making; Jane Pooley and her courageous outing on national television; Luis Elizondo, Eric Davis, Chris Mellon, Alain Juillet, and voices for the Pentagon who have gone on the record regarding official engagement with UFO phenomena; Mauro Biglino and Maxim Makukov for their courageous commitment to detail, cost whatever it might; experiencers like Patricia from Massachusetts and Dean from New South Wales for entrusting me with their stories; researchers like Richard Dolan, Erich von Daniken and my collaborator in *The 5th Kind TV*, Anthony Barrett, who have been bold enough to nail their colours to the mast on the most controversial of topics; and clinical psychologists like Barbara Lamb and the late John Mack who have honoured the courage of their clients with a bravery of their own. Their therapeutic work has helped to expose a phenomenon we should all be concerned about. Their amazing work has extended a helping hand to many people isolated by experiences of close encounters and has helped their scars to heal.

Fifteen minutes to go.

I look down at the marks on my right ankle left by the portable traction device I had to wear after my accident and smile at the memory of it. I feel grateful for the ultimate frisbee injury which laid me up and locked me down for weeks on end to begin the research that led me to *Escaping from Eden*. For me it was a happy synchronicity. The discoveries that followed opened up a brave new world for me, and the subsequent journey to *The Scars of Eden* has widened my perspective still further concerning our origins as a species and our potential as human beings. Today I have a renewed appetite as I look to the future and a more open heart to listen to others around me whose experiences have led them on to unfamiliar ground.

Ten minutes to go.

My phone buzzes. It's Jason from Mount Washington, confirming a coaching appointment. He is an army veteran, an engineer and a scholar of world history. He has a story to tell about his service overseas, and something unusual that happened in his childhood. Browsing on *YouTube*, Jason has stumbled across the *Paul Wallis Channel*, which has led him in turn to *The 5th Kind TV* and to one of our biggest documentaries – one about the Sumerian cuneiforms and the fascinating secrets they hold. Evidently, Jason was blown away by it. He has written me an amazing and lengthy message and is eager to share his experiences with me.

Five minutes to go.

I set my phone to silent and tap out an initial reply. *"Thanks, Jason, for your message. I would love to talk with you..."*

As people make contact with me from week to week, I am becoming more and more aware of the proportion of people who silently carry stories like Jason's. As the world moves further into the C21st, people are remembering. Still more people are taking an interest in what has been remembered, both by our contemporaries and by our distant ancestors. I am continually amazed by who I meet on this ET territory. Some get here through a journey in ministry, others through journeys in archaeology, paleobiology, mythology, literature, anthropology, psychology, or neurology. In fact, there would seem to be any number of red pills available to those who are ready to acknowledge the anomalies and follow the white rabbits where they lead.

Today I am sensing a rising tide of discovery and awakening. I am excited to see this rise of popular interest, because as we compare notes along the way, we give one another courage for the journey. I long to see a recovery of lost memories surrounding our origins as a species, because it is as we remember who we are and where we came from that we awaken to our true potential. The time has surely come to break the taboos, end the official embargos, and allow the whole of humanity on to the same page. I feel there is a momentum building.

Half a century ago, only a few short hours before the Saturn V rockets of Apollo 14 launched Ed Mitchell into space on his journey to the moon and back, he and his crewmates, Alan Shepard and Stuart Roosa, received a private briefing from the revered astrophysicist, Carl Sagan. A mutual friend tells me that, among other things, the briefing included the secret codeword for the three astronauts to use on this mission, in the event that they might find themselves accompanied by extraterrestrials on their way to or from the moon, or while on the lunar surface. Evidently, fifteen years into our space missions and on the third successful lunar mission it was clearly understood that a codeword would be needed, and Carl Sagan supplied it. In this way NASA could be fully apprised of an ET situation without the public ever needing to know anything.

Three decades later, Ed Mitchell was emphatic. His heart's longing was for disclosure. *"It is now time,"* he said, *"to put away this embargo of truth about the alien presence... now that we are space-farers on our planet... because we're really universal beings... We're at a point where we have to become part of a neighbourhood of inhabited planets... [something which] we have not acknowledged... exists – until this point! ... It is my hope that we can open this whole issue up."*

I feel the same.

In the last few years we have all seen our world change and perhaps we have a better idea than ever how disempowering it can be to be left in the dark with regard to vital information. The pattern of one truth for the privileged and one for the rest of us is not one I find acceptable. This is what the movement for official disclosure is all about.

Yet, with or without official disclosures concerning ET contact, past or present, if regular people like you and me can overcome our embarrassment, break the taboo and talk about this topic among ourselves, then our own networks of friends and family can become the agency of an unstoppable tidal wave of grassroots disclosure. The story is already here among us. It is in our past and in our present. As I look ahead, I am eager for the day when the human race can enjoy the benefits of all that our ancestors have memorialised and embedded so carefully in our world of mythologies. And I want to know the truth for myself.

Today as I sit in the comfort of Barbara Lamb's waiting room, I am feeling ready to delve more deeply into the recesses of my own mind and recover whatever lost memories may lie there, waiting to resurface. And if those memories give me even more to think about and upend conclusions that I thought were in the bag, then so be it.

The room is comfortable and bright with a nice flow of fresh air to keep me awake in the warm afternoon sun. What memories bring people into this room? What stories surface here? What scars are revealed? And what journeys of healing follow? I tuck my phone back into my jacket pocket and take a deep breath. That's the sound of the door from down the hallway. As it opens, I can hear Barbara's voice and the warm muffle of friendly conversation. I glance up at the clock. It's one minute to. I breathe again, a little deeper. Alright. I'm ready to go in.

**6TH
BOOKS**

ALL THINGS PARANORMAL

Investigations, explanations and deliberations on the paranormal,
supernatural, explainable or unexplainable. 6th Books seeks to
give answers while nourishing the soul: whether making use of the
scientific model or anecdotal and fun, but always
beautifully written.
Titles cover everything within parapsychology: how to, lifestyles,
alternative medicine, beliefs, myths and theories.
If you have enjoyed this book, why not tell other readers by
posting a review on your preferred book site?

Recent bestsellers from 6th Books are:

The Afterlife Unveiled
What the Dead Are Telling us About Their World!
Stafford Betty

What happens after we die? Spirits speaking through mediums know, and they want us to know. This book unveils their world…
Paperback: 978-1-84694-496-3 ebook: 978-1-84694-926-5

Spirit Release
Sue Allen

A guide to psychic attack, curses, witchcraft, spirit attachment, possession, soul retrieval, haunting, deliverance, exorcism and more, as taught at the College of Psychic Studies.
Paperback: 978-1-84694-033-0 ebook: 978-1-84694-651-6

I'm Still With You
True Stories of Healing Grief Through Spirit Communication
Carole J. Obley

A series of after-death spirit communications which uplift, comfort and heal, and show how love helps us grieve.
Paperback: 978-1-84694-107-8 ebook: 978-1-84694-639-4

Less Incomplete
A Guide to Experiencing the Human Condition Beyond the Physical Body
Sandie Gustus

Based on 40 years of scientific research, this book is a dynamic guide to understanding life beyond the physical body.
Paperback: 978-1-84694-351-5 ebook: 978-1-84694-892-3

Advanced Psychic Development
Becky Walsh
Learn how to practise as a professional, contemporary spiritual medium.
Paperback: 978-1-84694-062-0 ebook: 978-1-78099-941-8

Astral Projection Made Easy
and overcoming the fear of death
Stephanie June Sorrell
From the popular Made Easy series, *Astral Projection Made Easy* helps to eliminate the fear of death, through discussion of life beyond the physical body.
Paperback: 978-1-84694-611-0 ebook: 978-1-78099-225-9

The Miracle Workers Handbook
Seven Levels of Power and Manifestation of the Virgin Mary
Sherrie Dillard
Learn how to invoke the Virgin Mary's presence, communicate with her, receive her grace and miracles and become a miracle worker.
Paperback: 978-1-84694-920-3 ebook: 978-1-84694-921-0

Divine Guidance
The Answers You Need to Make Miracles
Stephanie J. King
Ask any question and the answer will be presented, like a direct line to higher realms... *Divine Guidance* helps you to regain control over your own journey through life.
Paperback: 978-1-78099-794-0 ebook: 978-1-78099-793-3

The End of Death
How Near-Death Experiences Prove the Afterlife
Admir Serrano
A compelling examination of the phenomena of Near-Death Experiences.
Paperback: 978-1-78279-233-8 ebook: 978-1-78279-232-1

The Psychic & Spiritual Awareness Manual
A Guide to DIY Enlightenment
Kevin West
Discover practical ways of empowering yourself by unlocking your psychic awareness, through the Spiritualist and New Age approach.
Paperback: 978-1-78279-397-7 ebook: 978-1-78279-396-0

An Angels' Guide to Working with the Power of Light
Laura Newbury
Discovering her ability to communicate with angels, Laura Newbury records her inspirational messages of guidance and answers to universal questions.
Paperback: 978-1-84694-908-1 ebook: 978-1-84694-909-8

The Audible Life Stream
Ancient Secret of Dying While Living
Alistair Conwell
The secret to unlocking your purpose in life is to solve the mystery of death, while still living.
Paperback: 978-1-84694-329-4 ebook: 978-1-78535-297-3

Beyond Photography
Encounters with Orbs, Angels and Mysterious Light Forms!
John Pickering, Katie Hall
Orbs have been appearing all over the world in recent years. This is the personal account of one couple's experience of this new phenomenon.
Paperback: 978-1-90504-790-1

Blissfully Dead
Life Lessons from the Other Side
Melita Harvey
The spirit of Janelle, a former actress, takes the reader on a fascinating and insightful journey from the mind to the heart.
Paperback: 978-1-78535-078-8 ebook: 978-1-78535-079-5

Does It Rain in Other Dimensions?
A True Story of Alien Encounters
Mike Oram
We have neighbors in the universe. This book describes one man's experience of communicating with other-dimensional and extra-terrestrial beings over a 50-year period.
Paperback: 978-1-84694-054-5

Electronic Voices: Contact with Another Dimension?
Anabela Mourato Cardoso
Career diplomat and experimenter Dr Anabela Cardoso covers the latest research into Instrumental Transcommunication and Electronic Voice Phenomena.
Paperback: 978-1-84694-363-8

The Hidden Secrets of a Modern Seer
Cher Chevalier
An account of near death experiences, psychic battles between good and evil, multidimensional experiences and Demons and Angelic Helpers.
Paperback: 978-1-84694-307-2 ebook: 978-1-78099-058-3

Haunted: Horror of Haverfordwest
G. L. Davies
Blissful beginnings for a young couple turn into a nightmare after purchasing their dream home in Wales in 1989. Their love and their resolve are torn apart by an indescribable entity that pushes paranormal activity to the limit. Dare you step Inside?
Paperback: 978-1-78535-843-2   ebook: 978-1-78535-844-9

Raising Faith
A true story of raising a child psychic-medium
Claire Waters
One family's extraordinary experience learning about their young daughter's ability to communicate with spirits, and inspirational lessons learned on their journey so far.
Paperback: 978-1-78535-870-8 ebook: 978-1-78535-871-5

Readers of ebooks can buy or view any of these bestsellers by clicking on the live link in the title. Most titles are published in paperback and as an ebook. Paperbacks are available in traditional bookshops. Both print and ebook formats are available online.
Find more titles and sign up to our readers' newsletter at http://www.johnhuntpublishing.com/mind-body-spirit.
Follow us on Facebook at https://www.facebook.com/OBooks and Twitter at https://twitter.com/obooks.